Holley Bishop has been a beekeeper for six years and has spent thousands of hours talking to and observing her bees, harvesting and tasting honey, and amassing a collection of related books, gadgets and stories. A graduate of Brown University, Ms. Bishop worked in publishing houses in Boston and New York and then at a New York literary agency. In 1999 she enrolled in the part-time program of the Columbia University School of Journalism, which she completed in 2002.

Praise for *Robbing the Bees:*

'Written with grace and wit, *Robbing the Bees* is as seductive as an open jar of Tupelo honey' Robert Michael Pyle

'Holley Bishop's love affair with honeybees combines natural and social history with gastronomy and memoir to produce a delicious reading experience' Michael Pollan

'Holley Bishop has gathered rich nectar from the pages of history and dribbled it around a fascinating portrait of a modern-day beekeeper . . . She has found real magic in the world, and the result is delicious' Trevor Corson

'How sweet it is! With a fascinating mix of biology, history, and business, *Robbing the Bees* is a savory reading treat. World figures as different as Aristotle, Martha Washington, and Winnie the Pooh have all succumbed to the power of honey. Read this book and you will too' Mark Obmascik

Holley Bishop

Robbing

the Bees

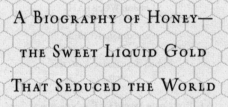

❖

A Biography of Honey—
the Sweet Liquid Gold
That Seduced the World

❖

POCKET
BOOKS

LONDON • SYDNEY • NEW YORK • TORONTO

First published in Great Britain by
Simon & Schuster UK Ltd, 2006
This edition first published by Pocket Books, 2007
An imprint of Simon & Schuster UK Ltd
A CBS COMPANY

1 3 5 7 9 10 8 6 4 2

Simon & Schuster UK Ltd
Africa House
64–78 Kingsway
London WC2B 6AH

www.simonsays.co.uk

Simon & Schuster Australia
Sydney

A CIP catalogue record for this book is available
from the British Library.

ISBN: 978-1-4165-1118-2

Printed and bound in Great Britain by
Cox & Wyman Ltd, Reading, Berks

Book design by Laura Lindgren

Dedicated to the original Rubygoo

Robbing the Bees

Contents

Introductions

For he on honey-dew hath fed,
And drunk the milk of Paradise.
Samuel Taylor Coleridge, *Kubla Khan*

Everyone should have two or three hives of bees.
Bees are easier to keep than a dog or a cat.
They are more interesting than gerbils.
Sue Hubbell, *A Book of Bees*

Nobody disputes the role of dogs as man's best friend, but a
convincing argument can also be made for the honey bee.
Martin Elkort, *The Secret Life of Food*

Until six years ago, I had no acquaintance with bees or honey. No childhood memories of painful stings while playing in the yard or climbing a tree, nor neighborhood friends who could boast of such a dramatic experience. There were no eccentric suburban beekeepers to spy on in my early days, no busy oozing tree nests, and never an ounce of honey in the kitchen of the house where I grew up. Preferring the Hardy boys and Nancy Drew to Winnie-the-Pooh, I had not learned to appreciate bees or honey. Bees were a vague, somewhat menacing presence, like malarial mosquitoes or the bogeyman. I had never personally met any and was perfectly happy to keep it that way.

Then, as a harried adult in need of a peaceful getaway, I bought a house in Connecticut two hours north of my cramped rental apartment in New York City. I fell in love with the landscape and solitude of the country, with the shade of giant maples instead of skyscrapers, and with the sounds of woodpeckers and doves waking me in the morning rather than the roar and honk of traffic. The house is over two hundred years old, a quaint brown clapboard colonial, rich in history and nature, which I set out to explore. Soon, I learned that my little haven, with its steep woods, rocky ledges, and spring-fed cattailed pond, had once been a tobacco farm.

Giddy with fresh air and a pioneering do-it-yourself fever, I fantasized about becoming some kind of farmer myself. I toyed with visions of a giant vegetable garden, an orchard, and a produce stand. I thought about acquiring sheep and making cheese and sweaters. Somewhere I read that one acre of grazing land can support one dairy cow and did the math on an unlikely herd of cattle. In the midst of my very improbable farm dreams (this was, after all, a part-time project, and I am essentially lazy), I went to visit my friend Ace, an expert in part-time projects. He introduced me to two white boxes of bees he kept in a meadow near his house. Immediately I was captivated by the idea of low-maintenance farm stock that did the farming for you and didn't need to be walked, milked, or brushed. The amount of gear and gadgets involved also appealed.

Ace handed me a plastic bear full of his most recent harvest, and when I tilted it to my mouth, head back, eyes closed, I really experienced honey for the first time, standing next to its creators. In that glistening dollop, I could taste the sun and the water in his pond, the metallic minerals of the soil, and the tang of the golden-rod and the wildflowers blooming around the meadow. The present golden-green moment was sweetly and perfectly distilled in my mouth. When I opened my eyes, tree branches and blossoms were suddenly swimming and swaying with bees that I had somehow not noticed before. Bees hopped around blooms in a delicate looping minuet. Determined to have sweet drops of honey and nature on my tongue on a more regular basis, I resolved to host bees on my own property. Keeping bees was clearly the most exquisite way to learn about my land, farm it, and taste its liquid fruits. As visions of sheep and cows faded away, I dropped my head back again and opened my mouth for more honey. That is how my love affair with bees and their magical produce began.

Like most love affairs, it quickly got obsessive. I started to see bees and honey everywhere, and everything reminded me of them. Honey suddenly appeared in every aisle of my supermarket and in the bubbles of my bath. The condiment packets at Starbucks were love letters from the hive. In the city, I saw "Busy Bee" courier services, "Bee-Line" moving companies, and bees dancing about the flowers of the medians on Park Avenue. When the initial infatuation had worn off, I did a little background check. Reading everything I could on beekeeping and bees, I became a little more enamored with every detail I uncovered about this humble creature's illustrious past. Most of the books I found on the subject were dated and musty, but their sense of fascination, which I now shared, was fresh and timeless.

○ ○ ○

Reverence for the bee is as old as humanity. Bees, in fact, were on this planet long before humanity existed. Ancient civilizations believed that bees were divine messengers of the gods, or deities themselves. Kings and queens of the Nile carved symbols of them into their royal seals, and the Greeks of Ephesus minted coins with their images. Emperor Napoleon embroidered the mighty bee into his coat of arms as an emblem of power, immortality, and resurrection. One day at the New York Public Library, while I was researching bees, one of my subjects blithely and loudly explored the reading room, causing widespread consternation. I felt thrilled by this visitation from the gods.

Honey was humanity's only sweetener for centuries, and historically seekers had gone to great and painful lengths to obtain their sweet liquid grail. It seemed to me, as I observed our often unnatural world of modern conveniences and sugar substitutes, that bees and honey, like poetry and mystery, had become sadly neglected and unappreciated. I had taken them for granted myself, but no more. I read dozens of journals and books about the bee, enough to realize that I was just beginning to grasp her vast repertoire of marvels. The glob of precious honey that I had poured into my mouth at Ace's was the life's work of hundreds of bees, a unique floral ode collected from thousands of blossoms in a poetic foraging ritual that has not changed in millions of years. Honeybees are mostly female; they communicate by dancing; and collectively they travel thousands of miles to produce a single communal pound of honey. They live for only several weeks and heroically die after delivering their dreaded, venomous sting. Bees shape the very landscape in which we all live by cross-pollinating and changing the

plants that nourish them. After decades of living in honeyless ignorance I added these divine insects and their delicious produce to my
recommended daily allowance of magic and wonder.

A few years later, having acquired my own bees and harvested
their honey, the love affair was still going strong (although it had
had its painful moments), and I decided to write a book about it, a
tribute to bees and honey that I hoped would convey the magic of the
hives and the timelessness and wonder of drizzling a bit of honey
onto your tongue. Because I was a hobbyist puttering around just a
couple hives and beekeeping is so much more than a hobby, I wanted
to find a professional beekeeper to tell part of the story, someone
with years of expertise and annual rivers of honey compared to my
weekend trickle. The story needed a guide much more experienced
than myself.

To find my sage, I went to one of my early research haunts, the
Web site of the National Honey Board. It has what it calls a honey
locator, a directory by state of commercial beekeepers and the types
of honey they produce. Florida and California were my first choices,
because they had the largest populations of bees and because I
wanted to see how bees behave somewhere different and warm. I
e-mailed a bunch of beekeepers in those two states explaining my
project and asking if I could come and spend a few days watching
their operation. Of the twenty solicited, Donald Smiley was the only
one who replied, from a place I'd never heard of: Wewahitchka,
Florida. In retrospect, I know this was because beekeepers are
extremely busy and hardworking, and writers from New York are
generally considered a nuisance. But Smiley alone took the risk
and the time and endured my endless questions because he is as eager

to celebrate bees and honey as I am. His honey epiphany occurred seventeen years ago and is still driving him with passion and wonder. "Hello, Holley," he wrote the day after my first e-mail. "Yes, I would be interested in helping you with the research for your book. The end of March may not be the best time for me though, the second week of April would probably be better. That is when our tupelo bloom begins, then it is all work and no play. Please give me a call and let's discuss it. The best time to reach me would be early morning between 5 A.M. and 7 A.M." In the first five minutes of our very early inaugural phone conversation he talked about his job with energetic wonder, joy, and pride and said, "I know I'm going to do this for the rest of my life." My thoughts exactly.

Note: There are an estimated sixteen thousand species of bees inhabiting our planet. From the stingless bees of the tropics to the giant honeybees of Southeast Asia, each has a distinct character and a fascinating history. This particular book is concerned with the genus *Apis*, which currently includes eight species of honeybees, the best known and most widely distributed of which is *Apis mellifera*, the Western honey bee. Within *mellifera* are twenty-four distinct races. I have focused mostly on the Italian race, *ligustica*, because I know it best. I keep *ligustica* in my own backyard, and Smiley too has long been smitten with it.

Another note: I visited Donald Smiley and his ever-expanding, ever-changing operation many times over the course of three years. Every time I arrived, there were more hives, new equipment, and usually a new assistant or two. When I first met him, Smiley had about six hundred hives; he now has well over a thousand. For clarity, simplicity, and sanity, I picked a number of hives, seven hundred (which is about what he had in the second year I visited), and made that constant throughout the story. Otherwise, I have gathered moments and events from throughout the three years that best illuminate a typical year in the life of Donald Smiley and his apiary.

ONE

Swamp Cache

"Where a man's treasure lies, there lies his heart."
Anonymous

Life is the flower for which love is the honey.
Victor Hugo

One year during the tupelo harvest Donald Smiley fell asleep while taking off his boots. He awoke slouched and snoring in his office chair, with one muddy boot on his foot and the other cradled in his lap. He's not to that point yet, he thinks now, in the middle of April, but give it a week. Seated in his swiveling desk chair, he pulls on one of the heavy, dark leather work boots, wincing a little at the effort. The season has barely begun, but already his 48-year-old body feels the strain. The most exhausting and lucrative time of year, the tupelo harvest, is fast approaching, and he's already tired from too much work and too little sleep. "Sometimes I'm so tired I can't get to sleep," he groans. "I stay up thinking about how tired I am and how much work there is to do." Easing a foot into the other boot, he achily ponders the work ahead.

He stretches and ambles into the adjacent kitchen to pour his third cup of morning coffee into a scarred plastic mug before heading out into the heat of the day. He likes it black, strong, and brewed to opacity. "That's what it takes to get me started," he says, drinking deeply as he swings the back door open. At 9:00 A.M. in April in the Florida panhandle, the temperature has already hit the hot, sticky seventies. In his frayed baseball cap, thick blue cotton work pants, undershirt, and long-sleeved denim shirt, Smiley breaks a sweat just by walking from his office to the truck parked next to the house.

Behind his beige-and-green ranch is a forty-foot-long steel Quonset hut as big as his residence. This is the honey house, the arched metal barn for storing his equipment and for processing his crop when it comes in from the fields. His livestock—about forty million bees—lives elsewhere, in twenty-two beeyards scattered around the swamps, fields, and forests of Wewahitchka. At the moment the honey house lies dormant, like a winery days before the grapes are harvested. But very soon the silent hut will be transformed into a bustling, fragrant honey factory.

In the adjacent sandy lot is another dormant structure, the Tyvecked shell of the two-bedroom house he is building for himself and his wife, Paula. His future abode is about 3,400 square feet, three and a half times as big as his current one, with a two-car garage, sunken bathtub, screened-in porch, office, and a separate storage space for honey. Paula, his wife of six years, is eager to move from his old bachelor pad and childhood home to a more spacious place that doesn't have boxes of honey stacked in the bedroom. She'll have to wait until the harvest is over. At the rate the contractors and bees are going (slow and fast), the Smileys probably won't move in until the fall. In the meantime, he has to pay for it. "Time to get to work," he says, swigging his coffee and eyeing the house. "I need the tupelo money to pay for the tiles."

He climbs into his gleaming white Ford 4×4, blasts the radio and air-conditioning, settles his coffee cup into its holder, and backs out of the driveway. From his house on Bozeman Circle, Smiley turns onto Old Transfer Road, then onto Route 71, the main drag intersecting the small town of Wewahitchka, or "Wewa," as the locals call it.

Geographically, Wewa is about equidistant from New Orleans, Birmingham, Atlanta, and Jacksonville. Culturally, Smiley's hometown is much closer to the old, deep coastal South of Louisiana and Alabama than it is to the sandy resort world of southeastern Florida. Locals often refer to this area as "L.A." or "Lower Alabama." Some call it "Southern Alabama," or "Salabama." The older houses in Salabama are built of wood, raised on wobbly stilts and roofed with tin. Ancient magnolia trees draped in clouds of Spanish moss dwarf and shade the single-story homes. Some lots have a trailer parked next to a crumbling older house, the local version of an addition. Trucks are parked on packed sandy earth by the side of the house, often sporting a gun rack in the rear and fishing poles poking out of the bed. In front of several homes on Route 71 are signs for boiled peanuts, honey, live shrimp, and crawfish for sale. From deep within shaded front porches, residents watch for customers and traffic on the road, which extends from the shore twenty miles south to the Alabama border fifty miles north. The difference between north and south Florida is, as Smiley says, that of night and day.

Most of Smiley's life in Wewa has centered around Route 71. He was born on it, in a house that used to be the doctor's office and is now Eddie's Beauty Salon. The elementary and high schools he attended are just down the road, as are his bank and supermarket. Even his mobile office—the truck—is headquartered on Route 71; he spends several hours a day on it as he tends to beeyards throughout the county. With a population of 1,700 people and only one main thoroughfare, Wewa is a town where strangers are noticed and where all the locals know all the locals, at least by sight. Smiley recognizes many of the drivers passing him and lifts his left fingers up from the steering wheel in a subtle, neighborly wave.

Stopped at the one traffic light, Smiley surveys "downtown" Wewa. Clustered near the intersection are Randy's discount grocery store, the Chevron station, Pitt's pharmacy, Tony's Bait and Tackle, JR's food mart and gas station, and an abandoned gas station turned failed doughnut shop. He can also see a hardware store, an auto parts store, a Subway sandwich outlet, several empty lots, and the Bank Trust of Florida. Farther up the road are the Lake Alice municipal park with its covered pavilion, pond, picnic tables, and a bright plastic jungle gym, and the new town library. As for other small-town entertainments, there is not a bowling alley, arcade, or movie theater in sight or in town. "Wewa used to have a movie theater," says Smiley, pointing. "Over by the where the courthouse is now, but it burned down before I was old enough to go to it." In the lifetime that Smiley has been here, lots of businesses have burned down or closed up, and the crossroads has the abandoned feeling of faded charm and better days found in many small southern towns. "We used to have a five-and-dime, and a shoe store and a clothing store, but they all went out of business," he says. "Had a dry cleaner too, but that closed." For entertainment, dry cleaning, cheap groceries, and home building supplies, Wewahitchkans drive to the thriving malls and restaurants of Panama City, about half an hour away.

Not counting the Subway, Wewahitchka boasts two restaurants, The Bayou and Swampy's. The menus are crowded with fried fish, okra, shrimp, and crawfish, with an occasional frog's legs special. It's the kind of food Smiley caught and cooked himself when he was a kid, crowded into the two-bedroom house on Bozeman Circle where he still lives. By local standards he's well off now, but that's only been in the past few years, since he started in honey. He and his twelve siblings were raised, as he says, "hard and fast and poor, same as most kids in this part of Florida."

Smiley drives past the high school he attended until eleventh grade, when he dropped out, married his sweetheart, and started work in construction and installing air-conditioning. He and most of his classmates stayed in the area, laboring in the local industries of farming, paper, timber, chemicals, and fishing. After the construction gig, Smiley tried timbering, working grueling shifts on an oil rig, and eventually guarding at a local prison. He preferred shifts in maximum security, because it kept him busy. "Regular duty you just sit on your butt and scratch your head and watch the prisoners," he says. Donald Smiley likes to be busy.

Early in his marriage, he and his wife and two small children moved thirty miles east to Apalachicola, Florida's oyster capital, where he worked plumbing the endless sandy shallows of the Gulf of Mexico for the prized shellfish. When a friend introduced him to his beekeeping hobby, it was love at first sight, sound, and taste. He remembers going into a hive, digging his chaffed oysterman's finger into the laden honeycomb, and tasting the warm waxy sweetness as bees swarmed noisily around him. It seemed an additional miracle that he might actually make money from this liquid. Phasing out of bivalves and into bees for a living, he returned to Wewahitchka and his childhood home in 1987. He bought eight hives from a newspaper advertisement and by the next year had turned them into forty. After years of teetering, his marriage finally tottered apart; for a year before he met Paula, he found himself living alone in his house for the first time in his life with only a couple of million bees to keep him company. Following his divorce, bees were his passion, his livelihood, his friends and family.

Now, after years of perfecting his husbandry, he has seven hundred hives and a new wife, whom he married after a six-month courtship. "I told her I was already married to my bees," says Smiley now, "but she married me anyway." Paula doesn't seem to mind sharing her man with the bees. In a crunch she helps him with sting-free tasks such as bottling, labeling, and shipping. She keeps a respectful distance from his beloved livestock but is clearly proud of his dedication and his growing reputation. In seventeen years, he's become one of the most respected beekeepers in the area. Other farmers and hobbyists come to him for advice. Customers ask for his honey by name. Smiley's work is hard, hot, and at times endless. But the rewards are sweet—financial and physical independence, a sense of craftsmanship, a constant education, and the joy of working outdoors, coaxing delectable honey from millions of bees. "I love my job," he says. "I made more money beekeeping than anything else I ever did and enjoyed it more." He's constantly amazed and impressed by the bees.

Easing up to the drive-through window of the Emerald Coast Federal Credit Union, he takes a sheaf of checks from his shirt pocket. These are payments for the honey that he shipped last week.

"Well, good morning," he says to the intercom, slipping the checks into the open transaction drawer.

"Hey, Mister Smiley, how's it going?" chirps the teller from behind the glass. "Did the tupelo come yet?"

"Well, we're fixin' to find out what's happening there any day now," replies Smiley.

The object of this speculation is a tree bud, or, more accurately, millions of tree buds. The tupelo tree, *Nyssa ogeche*, grows in profusion in and along the Apalachicola and Ochlocknee rivers of the Florida panhandle. A few of these trees are found in southern Alabama and Georgia, but they grow in significant numbers only in Smiley's little-known corner of Florida, commonly called the "Forgotten Coast." There are legends about how the tupelo, a native of China, came to be in this forgotten place. The most popular story involves a missionary, a Baptist woman who had just returned from the Far East and was traveling down the Apalachicola River by barge. In her purse she reportedly carried precious seedpods from a Chinese tupelo tree. It's not clear why she had these curious items in her bag; perhaps she had tasted tupelo honey in Asia (where beekeeping had been established for millennia) and intended to spread the word and some seeds throughout Florida's watery frontier. When a thief grabbed the lady's bag and saw nothing of value within, he disgustedly flung her parcel into the river. As it turns out, he did honey lovers a valuable favor. The tupelo thrives in riverine environments, and it flourished and multiplied and now dominates thousands of acres of Florida wetlands. This is how Don Smiley's hometown came to be one of the few places on the planet where tupelo honey is abundant enough to be produced commercially. There are several beekeeping families in town, including the Laniers and the Rishes, who have been chasing the tupelo bud for generations.

The crop begins the day the tupelo buds open. Honeybees feed on the nectar of the blossoms, which the plant produces for their delectation. They are attracted to the tupelo over all others and make a straight, determined "beeline" to the buds. At the sweet liquid center of each flower they fill their stomachs drop by drop with nectar. When their bellies are full, they return to the nest, where the contents are converted to honey and stored as food for the entire colony. As human food, tupelo honey is a delicacy prized for its exotic rarity, unique flavor, light color, and refusal to crystallize—as most other honeys do. It's a specialty honey that Smiley can sell for more per pound than any of his other harvests.

There are as many types of honey as there are flowers. Bees forage on whatever nectar source is closest and most appealing to the hive, and the type of bloom influences the flavor and color of the honey. A colony of bees is like a sponge, soaking up the pools of smell and taste from the flavorful landscape and season in which it is immersed. Hives placed in a grove of blossoming orange trees, for example, will yield a light amber-colored honey with a mild citrus tang and the aroma of oranges. A eucalyptus grove, on the other hand, will offer honey that is darkly aromatic and medicinal tasting. Leatherwood honey from Tasmania, with its exotic, spicy, aftershave-like scent, is my favorite outside of my own backyard product. Honey connoisseurs can detect provenance in the same way that wine aficionados can pinpoint grapes, terroir, and appellation. In my own small apiary, I can savor the difference between my first summer harvest, infused with the sweet clover, dandelion, and forsythia flavors of spring, and the later crops, which are dark and fragrant with fall-flowering goldenrod and sumac. To produce honeys boasting a particular flavor and color, beekeepers induce their bees to forage exclusively on one type or one mixture of nectars. They transport their colonies to the sources, surrounding them with flowers, encouraging their bees to soak up tupelo or citrus or leatherwood, depending on the desired taste and what is in blossom.

In Florida, fewer than ten species of plants contribute the bulk of the state's honey crop, and only one of them—citrus—is cultivated. Wild nectar providers include cabbage palm, gallberry, saw palmetto, black mangrove, clover, and, in the panhandle, tupelo. The tupelo season is short and intense, lasting for only two to three weeks a year, usually from mid-April to the first week in May at the latest. During these weeks, Smiley's bees must forage exclusively on tupelo nectar in order for him to harvest a pure and lucrative crop. Before and after the tupelo, Smiley moves his hives to clover, gallberry, titi, cotton, and watermelon to secure those less distinctive—and less profitable—drops of nectar. But at this moment, in the second week of April, he's focused on capturing the tupelo.

At Howard's Creek Road, Smiley turns onto tarmac bleached and cracked from the sun and cruises past tall pine groves and stumpy fields where the timber has recently been harvested. Where most motorists would see a grassy roadside bordered by generic scrub and shrub, Smiley sees bursts of flavor and income: two kinds of clover, some gallberry, titi bushes, and golden green acres of saw palmetto.

Eventually the road narrows and a canopy of silvery branches and leaves forms overhead. Smiley slows suddenly, as if a blinking neon road sign had demanded his attention. He brakes the truck and gets out, leaving the door open and the air-conditioning blasting as he strides to an overhead branch and pulls the tip of it down to inches from his blue eyes. At the end, surrounded by pointy oval lime-green leaves, is a trio of inch-long stems, each supporting a knobby bud the size and color of a fresh spring pea. Smiley inspects it closely, then lets the branch spring out of his hands. "Three days," he pronounces. Three days until the buds open to release a flow of tupelo nectar.

The size and color of the buds, today's heat, yesterday's rain, and years of experience figure into this prediction. If it is accurate, then in three days that bud will sprout tiny, sweet, white tendrils that will transform the hard ball into the soft fluff that is the tupelo flower. A delicate conspiracy of water and sun bring forth the fluff for just ten to twenty days each year. Smiley and the other commercial honey farmers in Wewa attempt to predict the beginning and duration of the tupelo nectar flow in order to maximize their harvest and their profits.

Smiley climbs back into the truck and continues down the sandy road, nearing the Apalachicola River, where the tupelo trees grow in dense profusion. He makes a left at a country store and eventually pulls up to a blackened wire fence hung with "Stop" and "Keep Out" signs. Vandals, both human and animal, are one of Smiley's prevailing beekeeping problems. Local kids with not much else to do like to kick over the hives and watch the bees storm out furiously. Bears also enjoy the sport and the sweet proceeds.

Smiley opens the gate at the Howard's Creek yard and pulls in the truck. He then backs it up so the bed is positioned at the beginning of two long rows of pale-painted wooden hives. The rows are fifteen feet apart, with about sixty boxes on each side, resembling trim, miniature town houses on a broad, grassy avenue. Thick, scratchy brown pine needles and pinecones the size of grapefruits litter the alley between the rows. This quaint village at Howard's Creek is home to six million bees.

He shuts off the engine and alights from the truck. The only noise in the beeyard now is from the breeze in the towering pines, an occasional cricket song, and the subtle, enveloping hum of millions of bees. Looking down the row of hives, Smiley sees them dipping and diving, turning from brown to reflective silver as they dart from shade to sun. He takes a deep breath to admire and analyze the floral scents in the yard. "It's a great day to be alive and keeping bees," he exhales.

Smiley takes a sip of coffee, reaches into the truck bed to find his wide-brimmed white plastic hat, and plops it on his head. On the down-turned gate of the truck, he places his smoker, a metal cylinder about the size of a coffee can with a conical spout on top and a bellows on the side. From the ground he grabs a handful of long brown pine needles, folding a thick sheaf of them in half. With his lighter, he ignites the bottom of this bundle and shoves it into the smoker while pumping the bellows. Sweet piney gray smoke billows out of the can, and Smiley adds a handful of wood chips from the back of the truck. While the fire grows, Smiley pulls the veil of his hat down over his face and anchors it with ties around his chest. In his rear pocket, he holsters a slender, black metal hive tool, a beekeeper's miniature crowbar. He flips the smoker closed, and smoke streams steadily from the spout.

Thus armed, Smiley saunters over to the first box, which is most often called a deep hive body and is the size and shape of a file box, complete with overhanging lid and incised carrying handles. These are man-made homes for the bees, who, as long as the space is neither too big nor too small for their numbers, keep house just as they would in the wild. Ruled by one fertile queen, each hive is a nest of wax rooms with eggs and brood at the core, surrounded by outer layers and walls of food supplies. Beekeepers in earlier times captured swarms in the wild and tended them in baskets, bottles, and boxes, but contemporary keepers actually breed and raise colonies of bees inside the wooden hives. In modern commercial hives, bees graft their nests onto a series of internal movable frames, which make it as easy for the beekeeper to monitor and manipulate his livestock as if pulling and reading a file from a drawer. Smiley's goal is briefly to inspect a couple of files or frames from each box to make sure that the establishment is healthy, that the queen is laying eggs, and that the colony is robust enough for the imminent nectar flow. He's also checking to see that they aren't too robust, threatening to outgrow and outnumber their space and swarm away in search of more suitable accommodation. Contrary to popular belief, a traveling swarm of bees is not a feral clump of stinging marauders but rather a vulnerable family of homeless bees that has outgrown its living quarters. Smiley is a good landlord, constantly checking to be sure his tenants are comfortable and adequately housed. He lifts the front of the lid about two inches to an emphatic burst of buzzing, greetings from a healthy, active colony. Wafting a few gray puffs into the opening, he sets the lid down again, giving the smoke a few seconds to sink in.

Various theories attempt to explain what the smoke does to a beehive. The most popular is that it alarms the inhabitants, which suspect a forest fire and then hide in the combs of the hive, binge-ing on honey until they become passive with sugar and fear. Some beekeepers say that smoke is a calming bee narcotic. My theory is that the bees simply want to get away from it, as if from bad breath, bad manners, or a bad cigar at a party. When they sense the obnox-ious, intrusive fumes, they withdraw politely and quietly into the private back rooms of the hive. Smiley believes the smoke merely disorients them and slows them down for a few minutes, which is all the time he needs. When he takes the lid off moments later, the bees have retreated and the hive is much quieter and calmer, hushed with smoke.

He taps the lid firmly on the hive to shake any wanderers back into the interior. Crawling bees coat the tops of the ten wooden frames that fill the box. "Hoo, boy. Lookee here. Now, there's a pretty box of bees," he croons. Lively dialogue with bees is as old as apiaries. In ancient Greece, where it was believed that honey rained down from the heavens above, bees had high status as collectors and dis-tributors of the nectar, and thus as liaisons to the gods. It was impor-tant to be on good terms with the bees if you needed anything from the deities. Throughout Europe until recently (and in some places still) bees were considered part of the family, and it was customary to inform them of important events, such as christenings, funer-als, and weddings. If you did not tell the bees (and share a piece of wedding cake with them), it is said, they were likely to leave you, and your life would no longer be sweetened with honey.

Superstitions and deities aside, Smiley enjoys talking to his bees. Like farm animals or pets, they respond to the tone, mood, and movements of their master. His relating the events of the day in a soothing voice and calm gestures seems to soothe them as much as it does him. Besides, he has no one else to chat with on the many days he spends alone in the beeyards.

Smiley squats down beside the box, the smoker puffing lazily beside him. With weathered bare hands, he wedges the hive tool in between the first frame and the lip of the box and levers out a segment of the hive. Bringing the 9 × 18 × 1-inch frame close to his veiled face, he peers at the contents like a doctor reading an X-ray. There are approximately 4,000 hexagonal wax cells on each side of the frame. Each six-sided cup is about a quarter inch across and a half inch deep. At the bottom of many of the cells are single white specks the size of poppy seeds: sausage-shaped eggs laid recently by the queen. Smiley is pleased at this sign of a busy, fertile monarch. When the eggs have matured, this frame will yield thousands of new bees, and the more bees he has, the more drops of precious nectar they'll bring back to the hive. "If you've got eggs, you've got honey," proclaims Smiley as he inspects the frame. Every little white speck counts.

On top of the heartiest colonies, those already teeming with bees and brood, he has placed another box called a honey super. It is shallower than the deep hive body but otherwise identical, filled with ten frames of honeycombs that the bees have manufactured using wax secreted from their bodies. Away from the queen and her burgeoning offspring, this is where the bees make and store their food. Between the two levels is a queen excluder, a thin metal wire rack whose bars are close enough together to allow workers access and to prohibit the slightly fatter queen from getting into the food bank and littering it with her relentless egg laying. While the lower level is the actual home and nest of the bees, the upper honey annex is an interchangeable storage area. This is the bee-keeper's harvest.

Lifting one of these heavy frames from a super, Smiley sees it is thick with glistening honey. Made from the nectar of tiny white gallberry flowers, its dark amber color earns it the nickname "red" from local farmers. As a rule, the darker a honey is, the stronger its flavor. Smiley scrapes some red from the comb with his index finger and lifts it beneath the veil and into his mouth, taking the power-ful floral explosion onto his tongue. He ponders the flavor like a sommelier, tasting a hint of black titi, maybe some clover and hon-eysuckle, lots of gallberry. The latter he senses as a hint of bitter-ness at the back of his throat. This is the taste of end-of-the-flow gallberry.

Because most plants flower for only a few weeks at most, and in different locales, honey farming is a surprisingly physical and migra-tory endeavor. Like shepherds guiding their flocks from bare, tram-pled fields to succulent green meadows, beekeepers must also follow the forage, moving the hives from one emerging source of nectar to the next. Smiley moves his at least two or three times a season—typically from gallberry to tupelo to cotton and watermelon—chas-ing liquid abundance, capturing different nectars, absorbing a medley of flavors.

Red honey is produced in March and April, when gallberry is flowering throughout the panhandle. Usually, just as those blos-soms fade midmonth, the tupelo flowers open, releasing their flow of distinctive nectar. When that onslaught comes, the real artistry and work begin. In the business of specialty honey, there can be no nectar overlap, no mixing to dilute the purity and profitability of the crop. Tupelo honey must be pure tupelo honey. In the next three days, Smiley has to remove each of the current supers and extract the contents, cleaning out every last taste of the red. Then, in the very early mornings or cool evenings, when the bees are all calmly inside their hives, he'll load the colonies onto his truck and move them to the tupelo yards. There, he'll place the freshly emptied supers atop each hive and wait for the bees to bring in the new harvest. While the trees flower, the bees will make tupelo honey, which is as distinctively light and mild as the red is strong and dark. When the tupelo blossoms wither, Smiley repeats the whole exhausting process, immediately extracting the tupelo honey, then moving all of the hives to new yards and new sources, adding yet another set of empty supers to tap those nectar flows.

From another deep box, he pulls out a frame full of puffy beige brood cells containing eggs that have hatched and thickened into pupae. As Smiley watches, the papery cap of one of the brood cells is pierced from the inside. Antennae appear, and seconds later two dark bulging eyes emerge. An adult-size newborn bee crawls out of the cell to join the busy traffic on the frame. The slightly paler fur on her thorax is the only clue of her recent birth. The newborn instinctively begins to work and soon blends in with her thousands of siblings. "That sure is a beautiful thing," Smiley says to the frame as he replaces it in the hive and refits the lid over the colony.

He repeats the inspection process down the line of hives. Puff the smoke, tap the lid, applaud the bees, admire the honey, scrutinize a frame or two. It's a quiet, meditative process carried out in a haze of heat and smoke, punctuated by occasional whoops and exhortations. On one frame he spots the golden queen, with her distinctive long, tapered abdomen, making her way from cell to cell, determinedly laying eggs. An entourage of helpers follows the monarch, feeding her, cleaning her, ushering her across the comb. Smiley waves them on, "Good job there, girly, you just keep up the good work."

In another hive, he spots another queen and whistles at the abundance of red honey her colony has stored. Bees are instinctive hoarders. As long as nectar is flowing, they will compulsively stock their pantry supers with honey. Their intent is to store food for the winter months, when supplies are scarce. When nectar is abundant, the inhabitants of a colony will collectively fly 55,000 miles and gather from more than two million flowers to make a pound of honey, with each bee contributing in total just a twelfth of a teaspoon to the communal coffer in her lifetime. Although one hive can make as much as 150 pounds of honey in a summer season, it takes about a third of that to sustain a hive through the winter months, depending on the length and depth of the cold. From the time of the earliest apiaries, the arrangement between beekeepers and their bees has been for the keepers to harvest or "rob" the surplus summer honey, leaving enough in the hive for fall and winter sustenance. Smiley's workers need very little surplus to get them through the short, mild winters, so almost all of their instinctive excess is for his benefit. In Connecticut, however, my bees require about sixty pounds of honey per colony to sustain them during the long cold nectarless winters. I gratefully leave them this allowance, after stealing nearly twice that much from their pantry.

Smiley moves slowly down the line, smoking, tapping, and inspecting, taking the pulse and flavor of each colony. Eventually he returns to the truck, stows his hat and smoker in back, wipes his sweaty forehead, and unbuttons his clinging wet shirt. It is just before eleven in the morning. Smiley climbs into the truck, takes a gulp of cold coffee, blasts the AC, and turns on the radio just in time for the weather report. Afternoon showers, a regular panhandle occurrence, are expected. He doesn't mind getting wet, but the bees don't like to be rained on, it makes them awkward and irritable, so he'll have to take it into consideration. Gazing up through the windshield, he notes the cloudless blue sky, checks the dashboard clock, reviews a mental map of his yards, and pinpoints the next stop. The smoker in the back of the truck puffs a steady gray banner as he pulls back out onto the road and heads to the next inspection.

The scent of the smoke lingers in the Howard's Creek yard after the truck has faded from view. Calmer rhythms return as the bees shake off the smoke and resume darting into neighboring woods, fields, and swamps in search of nectar for the colony. In the twenty minutes it takes Smiley to drive to his next yard, hundreds of dramas will unfold in each hive at Howard's Creek.

Female worker bees will travel up to two miles from the nest in search of nectar, pollen, and water supplies for the hive. That's more than 8,000 acres of perusal at the disposal of each colony. In carefully situated beeyards like Smiley's, food sources are much closer and the bees don't have to travel very far to get to work. When a bee spots and smells a likely flower, she lands on the blossom or any part of the nearby plant that will support her forty-milligram body weight. She unfurls her flexible proboscis from beneath her chin; like a tiny elephant's trunk, it searches out the nectar pools, then sucks until all of the liquid within its reach is taken up. Draining up to 1,500 nectaries in this way, a bee fills her stomach or "honey sac," collecting up to half her weight in nectar before returning heavily to the hive. If nectar is abundant, achieving a full load can take as little as fifteen minutes. She might also be in search of pollen, the plant protein that bees feed their young, which she collects in saddlebags on her rear legs, packs into pellets, and transports back to the colony. Or she may go in search of a drop or two of the five gallons of water it takes to hydrate and cool the colony each year. These various hive duties add up to five hundred miles of flight in the lifetime of an average bee, a round trip from Wewahitchka to Orlando as the insect flies.

Arriving back at the yard with her foraging spoils, the bee recognizes her own home by the unique pheromonal scent of her queen, an aromatic chemical coat of arms that is relayed from bee to bee and tracked around the hive. Each bee carries a bit of this scent with her, a perfumed security pass that the guards at the entrance identify before waving her through the door. Wrong-smelling trespassing robber bees are turned away, discouraged from pilfering honey that isn't theirs. At the threshold, a returning forager empties her stomach and relays her load of nectar to waiting young "house" bees, who dutifully move off to process the nectar, extending their long tongues to offer the droplet to the warm drying air of the hive. Soon, they offer the nectar to other worker bees, who deposit it in storage cells and fan their wings to dry it further. When the honey is fully cured, reduced from 80 percent to less than 17 percent water, or from the consistency of sugar water to that of molasses, the bees cap it with wax for storage. There are different sizes of super frames. A full seven-inch-deep one is a tasty seven-pound liquid slab of stored honey veneered in wax.

The bee that delivered the nectar will take a short break, perhaps snacking on honey, then head back out to forage again, making up to thirty such trips a day. If she has found a particularly good nectar source and needs reinforcements, she will dance. Bees perform precisely choreographed dances to communicate fear, alarm, joy, and the locations of food and water. In the joy dance, the forager places her front legs on the back of another available bee and shakes her abdomen in a kind of bee conga. The interior of the hive is dark and crowded, so the dance is more of a jubilant mosh pit, with participants communicating through touch, movement, smell, and sound. Other bees will mimic the mosh movements before flying out to the reported loot.

Honeybees live for about six weeks on this diet of nectar and hard physical labor. Throughout this lifetime, their contributions to the community are dictated by their age. From the moment they are hatched, young bees go to work as custodians, cleaning up their own larval debris from the brood cells. In their first twenty-one days, they remain within the hive, working as carpenters, guards, and nurses. They help feed and clean the queen, build new comb, nurse the larvae, cure honey, and take sentry shifts protecting the entryway. Some work on the air-conditioning system of the colony by fanning their wings at the entrance or distributing water to the warmer realms of the hive. Finally, young bees take brief orientation flights outside to familiarize themselves with the local landmarks and plant nectaries. At the ripe middle age of three weeks, the bees leave the nest to forage, and from then on their lives can be measured in distance. Five hundred miles of toil and flight take a heavy toll. With torn and tattered wings and exhausted bodies, they wear out and die after about twenty days outside. Always gracious, economical, and neat, worker bees usually take a final flight and expire away from the hive.

Hundreds of bees die on any given day, but just as many or

more are born to continue the endless work of the family. In the height of summer, up to 60,000 adult bees live in a single colony. About half of these are older foraging bees laboring in fields and flowers, while the other half are newer bees working inside the hive, tending to 35,000 young in various stages of development. Although the lives of the worker bees are fleeting and expendable, the queen lives for up to several years and is the beekeeper's (and the hive's) most important investment. She is often a mail-order bride, arriving in an overnight envelope stamped "livestock" on the front from a queen wholesaler. Within the envelope is a screened wooden cage containing the queen, a food supply, and an entourage of five or six helpers. Many honey farmers replace their royalty each or every other year, ordering fresh young queens by the case to ensure youthfully robust fertility.

The alternative to the mail-order method is for the colony to create its own queen. If her attendants sense that they need a new leader—because the old one is dead, unwell, or not laying in a proper royal manner—they can grow a replacement. To do this, they construct special wax cells around seven or eight existing fertilized eggs, creating oblong armored incubators that resemble small peanuts. The female eggs and larvae in these cells are slathered almost continuously with royal jelly, a vitamin-rich hormonal goo secreted by the worker bees. After about two weeks of this spa treatment, a new monarch emerges from one of the queen cells and goes directly to the other peanut cells to sting and kill her erstwhile competitors, who may have been only minutes behind her in the race to hatch and claim the crown. Then, as one of her first royal duties, the new queen often unceremoniously murders her poor ailing mother. Sometimes she lets the queen mum stay, stripped of her crown and relegated to a neglected corner of the hive.

The victorious queen enjoys her new status, virginally, for about six days. Then she embarks on her only adventure outside of the hive. Leaving the nest, she soars high into the air to mate. Male bees have been waiting for this opportunity, and as many as ten will have the aerial pleasure of ejaculating a total of three to eight million spermatozoa into the royal oviducts. Her daylong honeymoon complete, the queen then returns to the hive. As the only fertilized female, she will spend the rest of her life confined to the nest laying up to 1,500 eggs a day. Royalty's is not always a life of ease.

The queen determines the sex of her offspring by parceling out the sperm from her maiden flight. Fertilized eggs become female worker bees. Eggs that remain unfertilized become drones. Every colony is different, but in the active season most house between 500 and 1,000 drones and 50,000 to 60,000 workers, making the average hive about 99 percent female. This female population performs all of the work of the hive. Males, or drones, can't fly well, and they don't gather food, clean, sting, secrete wax, or care for the young. When not gorging on pilfered honey in the nest, the males occasionally make sorties to locations known as drone-congregating areas. If they detect the pheromones of a virgin queen, they pursue her. A few succeed in mating with her, but those lucky few die soon afterward, their barbed genitals having broken off during copulation.

Except for the (dead) stud drones, males are useless to the hive, especially in commercial apiaries, where queens arrive already fertilized. With a fertile monarch in residence, the female worker bees persecute the drones by withholding food and sometimes

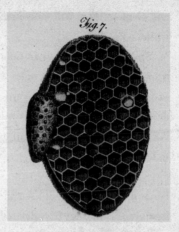

Drawings of the queen, drone, and worker from the 1889 edition of *Langstroth on the Hive and the Honey Bee.*

Figure seven (from a 1748 book) depicts a queen cell.

gnawing off their wings and legs in an effort to evict them. Most go willingly and die outside in a few days' time. Reluctant adult males and drone larvae are often dragged to the entrance of the hive and dramatically pushed out. By the fall, when nectar resources are scarce, no male freeloaders are left in the colony. The queen, abundantly fertile and dutifully laying her eggs, does not seem to care.

While Smiley drives to his second yard of the day, queens in the Howard's Creek hives will lay hundreds of eggs. A thousand sips of nectar will be converted to honey on tiny extended tongues. Bright orange and yellow loaves of pollen will be delivered to storage cells. Pupae will hatch, adult workers will die, and drones will be dragged to the door. Young bees will make their first flights from the hives, and older workers will do a dance and follow them out into the world.

Smiley's jaunt ends at an isolated beeyard situated fifteen feet from the brownish green water of the river. As he watches, bees exit the hives in the shady yard and travel into the swamp—no doubt, like him, checking to see if the tupelo nectar has arrived. For them it is an annual succulent feast; for him, a small fortune in honey.

For as long as beekeepers have understood that honey comes from flowering plants, they have devised ways to get as close to the source as possible. Ancient Egyptians, like Wewahitchkans, were accomplished beekeepers and built floating apiaries to collect nectar from up and down the river Nile. Centuries later, in 1740, the French traveler Benoist de Maillet noted that in October the beekeepers of Lower Egypt sent their hives on barges up the Nile to where it was warmer and the seasons changed sooner. As the procession floated downriver, bees were released from their hives to collect the nectar nearby. When the flowers in those pastures were exhausted, the barges were moved a few miles farther south, keeping up with the nectar flow. By the time the rafts arrived back in Cairo in the beginning of February, they had boatloads of honey for sale in the urban market.

Pliny the Elder, the first-century Roman scholar and author of the massive *Natural History*, observed similar practices in northern Italy and Spain:

Hostilia is a village on the River Padus. When their food supply fails in this region, the local people put the hives on boats and

carry them 5 miles upriver by night. At dawn the bees come out,
feed, and return every day to the boats, whose position alters until
such time as they have settled low in the water, under the very
weight of the honey—an indication that the hives are full. They
are then taken back to Hostilia and the honey is extracted. In
Spain the locals transport the hives about on mules for a similar
reason.

At about the same time, the Greeks were becoming connoisseurs of the nectar of the gods. They observed that the taste of honey varied depending on the source from which it predominantly came. One aficionado wrote: "Thyme yields honey with the best flavor: the next best are Greek savory, wild thyme and marjoram. In the third class, but still of high quality, are rosemary and Italian savory. Tamarisk and the jujube tree have only a mediocre flavor."

Most Greeks particularly liked the flavor of the honey from the thyme-covered slopes of Mount Hymettus, near Athens. Attic thyme honey was thought to be the most desirable in ancient Greece, and beekeepers competed for the right to maintain hives on the mountain. There are said to have been twenty thousand stocks of bees there at the time of Pericles—about 400 B.C. The mountain was so crowded with hives and itinerant beekeepers that Solon, the great Athenian legislator, passed a law mandating a distance of 300 feet between one set of hives and the next. Plato bemoaned the deterioration of the land by excessive beekeeping, referring to "mountains in Attica which can now support nothing but bees."

From Attica to Wewahitchka, specialty honeys have been produced through savvy, strategic placement of hives. On the Apalachicola until about twenty years ago, barges and permanent hive platforms were frequently used to gain access to the abundance within the swamp. When the tupelo flow started, beekeepers loaded hives onto boats and took them deep into the swamp or ferried them out to wooden river platforms on stilts that could hold hundreds of colonies. From these movable hive cities, the bees gorged on the tupelo forest. In 1878, a Chicago honey dealer built a barge to reap the nectar up and down the Mississippi River. The *American Bee Journal* described the floating bee metropolis:

> *The hives stand in four walls, five hives one above the other, nearly the whole length of the boat, about 250 hives in each line. The walls of colonies on the right side and the left side have openings for the bees to come out on the water front; a space of two feet between the hives and the guards answers for a gallery and for the bee man to walk on in front of the hives. In the middle of the boat there are two walls of colonies, 250 hives in each, facing an inner court six feet in width. The bees from these colonies reach the open air through the sky line opening in the roof above the court. Between the first and second rows of hives from the outside there is an aisle three feet in width, for the convenience of handling the hives and the honey. The distance from the barge deck to the roof over the colonies is fifteen feet. The space below the deck is ten feet in width and about seven feet high, and is to be used for sleeping apartments, making and repairing hives, handling and extracting honey and putting it in marketable shape. The dining room and cooking will be on the steamer that tows the bee fleet.*

In Britain in the fifteenth and sixteenth centuries, flowering heather was a preferred honey source. Records show that those

A riverboat headed south to Florida in the 1920s. Its load of
beehives can be seen stacked at the end of the ramp.

who lived "within perambulation" of the heather were allowed to
keep hives there. In 1786, T. Wildman wrote in his book *On Shifting
the Abodes of Bees*, "If there is heath at a convenient distance, the
hives being carried thither would considerably lengthen the season
of collecting this honey." Hives at that time would have been transported on mules, carts, specially adapted wheelbarrows, or the beekeeper's back. Cars and trucks have made the modern migration
considerably easier.

Maps from Wildman's day indicate several "bee gardens," or
encampments of hives, near water, sun, heather, and other choice
flowers and trees. These gardens were often in large estates or royal
hunting preserves, where the building of walls and enclosures was
strictly forbidden, so, like prospectors staking a valuable claim,

beekeepers frequently camped out next to their hives to secure the location and the nectar flow and protect it from competitors and predators. In Austria in 1760, the Empress Maria Theresa, a fan of all things bee-related, enacted a law forbidding any molestation of "temporary apiaries" or of the beekeepers camped nearby.

Honey enthusiasts did not always have to chase, barge to, or camp out by the nectar. Often they brought bees and sources closer to home, placing hives in domestic gardens and orchards, where the plantings could be varied in order to please their bees and their own palates. The Roman author Varro wrote of two brothers whose father had left them "only a small villa and a bit of land.... They had built an apiary entirely around the villa, and kept a garden." Varro then lists the flowering plants that the brothers raised in the garden in order to produce honey to their taste. In his 1618 book *A New Orchard and Garden*, William Lawson of London suggested placing hives next to lavender plants in the garden so as to produce "a most-fine lavender-flavored honey."

Honeybees, like the tupelo of which they are so fond, are not native to North America. They were introduced to the colonies by Lawson's countrymen and by Spanish missionaries early in the seventeenth century. There were probably bees on the *Mayflower* when it arrived, although there is no record of it, but a year later, in 1621, the Council of Virginia Company in London wrote to the governor of Virginia of another vessel: "We have by this ship... sent you divers sorte of seed and fruit trees, as also Pidgeons, connies (rabbits), Peacocks and beehives, as you shall by the invoice perceive: the preservation and encrease whereof we recommend to you." Native North Americans at that time had never seen bees or honey and had no words for them. John Elliot, the New England

Puritan pastor who was translating the Bible into native dialects for the Algonquin and Cherokee tribes (in hopes of converting them), is credited with giving them some, describing bees as "white man's flies." Whether the Indians converted to Christianity or not, they soon became devotees of the white man's flies and the delicious liquid they produced. To the south, in Central America and Mexico, natives had been enjoying honey from a local species of stingless bee for centuries, but that weaponless wonder had stayed in the comfort of the tropics, leaving the colder, uncomfortable North to be populated by the European man's flies. Colonies of these bees migrated gradually south along with the settlers, and by 1763, William Bartram reported that bees had been imported to Pensacola (about an hour from Wewahitchka) by the English, who took possession of Florida from the Spanish that year.

In Europe and later in North America, large landowners did not rob the bees themselves but allowed others to do it on payment of certain amounts of honey and wax. Like these beekeepers before him, Smiley leases his yards and offers honey as payment. Most of Smiley's lessors are grateful for the honey and the valuable cross-pollination that comes with the arrangement. Raw local honey is excellent currency, whatever the transaction. When a repairman came to fix my garage door so that I could get to my beekeeping equipment, he suggested a reduced fee in cash combined with a bottle of honey as payment in full. Smiley secures some of his locations with additional cash payments, guaranteeing that he gets his preferred sites every year. Yards that are secluded yet accessible, close to the tupelo trees, and reasonably close to town are at a premium. Local beekeepers know where the best locations are, who owns them, and who leases them. They swap stories and statistics about the yards as if they were discussing athletes and batting averages. Smiley is standing in his star player, his best and most expensive yard, which he leases for $600 and a gallon of tupelo a year. On this 600-square-foot patch of prime waterfront real estate, his bees camp and forage in the midst of thousands of tupelo trees. From the sixty-four hives placed here for three weeks of a good flow, he could harvest about eight barrels, or eight to nine thousand dollars' worth, of tupelo honey.

As he surveys the yard, calmly obsessing about his bees, he is calculating tupelo buds, hive yields, transportation times, and, as always, the weather. If the weather doesn't cooperate, his harvest could be halved. "Too much rain can end a tupelo flow in just a few minutes," Smiley observes, pulling pensively on his thin mustache. Too little rain can have the same effect. It has been a very dry spring so far, and as a result fewer flowers have blossomed, and have blossomed later, reducing the forage for his workers. Drier conditions also mean less sugar in the nectar and a less flavorful honey. When it's very hot, as it has been, the bees lounge around the hive, rallying to work only in the cool mornings and late afternoons. "When it's too hot, those bees don't hardly work at all," says Smiley with a shrug. "And they eat what they take." After two years of perfect conditions and bumper tupelo crops, this one doesn't look promising. The next few days will tell. He never knows for sure until the blossoms open and the bees bring home their first sweet report.

Overhead, dark clouds gather as Smiley inspects the tupelo branches, spying the same hard green buds he saw on Howard's Creek Road. The forecast here looks about the same as it did back there: three days. In the next seventy-two hours he has to finish inspecting all the current yards, extract many tons of red honey, move 700 mature colonies to the dozen or so tupelo yards, mend some hives, clean the honey house, and maybe get a few hours of sleep. He'll probably dream about bees, since during the harvest season, he regularly dreams about bees and honey. Visions of oysters, timber, or criminals from his other jobs never made it into his slumber, but for years now he's been dreaming about bees as if tending them in his sleep.

He dons his veil and does a quick hive inspection, using the smoke left over from Howard's Creek. These colonies look right on schedule, ripe with bees and ready for the flow. This is the calm before the tupelo. He takes a moment to enjoy the low lullaby hum of the beeyard before climbing back into the truck. From the budding treetops to the white boxes to the mirrored calm of the swamp, he gives the tableau one final admiring glance, then heads home for lunch. He's daydreaming about his bees and a magnificent harvest as the first drops of rain splatter against the windshield.

TWO

The Bee Master

*"The bee master must be first of all a bee lover,
or he will never succeed."*
Ticknor Edwardes, *The Lore of the Honey Bee*

The first time Smiley harvested honey was a disaster. In the summer of 1989 he had forty boxes ready to harvest, but he knew nothing much about bees except, as he says, that "they stung and made honey." Even so, he was looking forward, somewhat nervously, to his first crop. The first challenge was to get the honey supers away from the bees so he could take their loot elsewhere to extract it. Geared up and sweating in his hot new suit, he put some borrowed fume boards on the first bunch of hives. Treated with a noxious-smelling acid, the boards are intended to persuade the bees to flee the honey supers so their contents can be removed easily. But sometimes the bees need a lot of persuading. When he pulled the lids up a few minutes later, the laden frames within were still crawling with bees, probably, he realizes now, because the day wasn't hot enough to fully activate the fumes and chase the bees out of the honeyed regions of the hive. Not knowing any better, he loaded the teeming supers onto the bed of his truck, bees and all. Once on board they started stubbornly reclaiming their foodstuffs, and those left in the hives and yard joined in, so that the boxes, his truck, and his body were soon crawling with bees. "You couldn't even see the box for the bees. They were just boiling out," he remembers with amused embarrassment. "It seemed like every bee was on the outside of the box, or on me." By the end of the afternoon he had assembled a giant writhing swarm on the back of his truck. And it was getting dark.

He drove the swarm through town to the ten-by-ten-foot hand-hewn cypress-wood shack that he intended to use as a honey house. The ride was a rough one, dipping over dirt roads, with the hives sliding and shifting ominously in the truck bed. "I nearly killed 'em all on the ride over, I think," he laughs. By the end of the journey, the bees were furious. When he pulled up at the shack, they were practically growling as he wrestled a ferocious box into the house to begin extracting. Inside were a table, a wax uncapping knife, and the 32-frame extractor he had purchased from his friend Carl for $250. When he turned on the single light over the table, it was immediately covered with a layer of angry insects thick enough to eclipse the light. The lone bulb was a perfect beacon and target for all their displeasure. They flew at the light and the bewildered beekeeper, covering everything in a coat of angry bees. As the harvest progressed, he realized that his honey shack wasn't exactly bee-proof—bees were crawling through the gaps in the planking and even under the door to get at the man and to reclaim the honey inside. Carl kept a beeyard nearby, and his bees joined the twilit melee. The more Smiley "extracted," it seemed, the more bees gathered in the house. "There were more bees in the honey house than in the beeyard," he recalls with a sheepish smile.

Smiley managed to get the frames into and out of the extractor, sweating in his suit and wincing as he heard bees squish and crunch under his clumsy, nervous ministrations. On the bumpy return to the yard he couldn't see out the back window of the truck, caked as it was in bees. This was not the proud harvest he had imagined. "Looking back at that first year, I can't believe I'm still in the business," he says. "I knew there was probably a better way to do it. I just didn't know what it was. But I wasn't about to quit." He was flustered and confused but completely seduced. For centuries, men have been similarly beguiled, inspired to find a better way into the heart of the hive.

Smiley began by reading the beekeeper's bible, *The ABC and XYZ of Bee Culture*. Twice. Starting with "Abnormal Bees" and progressing to "Wladyslaw Zbikowski," the book is an alphabet of practical beekeeping questions first published in 1877 by Amos I. Root, who had fallen in love with bees dramatically and irrevocably, as most beekeepers do, a decade before. In August 1865, Root, who was then a jeweler in Ohio, passed a swarm of bees hanging from a tree bough. "My fellow workman in answer to some of my inquiries as to their habits, asked what I would give for them," he wrote in his journal. "I, not dreaming that he could call them down, offered him a dollar and he started after them. To my astonishment he, in a short time, returned with them hived in a rough wooden box he had hastily picked up, and, at that moment, I commenced my ABC in bee culture."

Newly smitten, Root spent hours observing his beloved, questioning, interpreting, and watching every move and mannerism of

An early advertisement for
Bee Culture Magazine and other
beekeeping products and services
from A. I. Root and company.

his darling. He conducted exhaustive backyard beehive experiments and shared his findings in *Gleanings in Bee Culture*, a magazine he started and that is still circulating today, a 135-year-old reference, forum, and fan club. In Illinois, Charles Dadant, an equally enthusiastic bee admirer, started another fan club, the *American Bee Journal*, which is also still in circulation. Throughout history, beekeepers (and writers) have been similarly seduced, compelled to discover, understand, and share everything they know about bees.

Roughly three quarters of a million different types of insects are now known to exist on this planet, and none has been more scrutinized, examined, or celebrated than the bee. Cave paintings, ancient temple walls, and hundreds of tomes and journals record man's curiosity and reverence for this creature. The obsession started in ancient Egypt, land of the earliest apiarists, who relied heavily on mystery and myth in trying to comprehend the life of the bee and the magical liquid associated with it. Because bees were seen gathering from the blossoms of upturned flowers, the Egyptians assumed they were collecting honey that had rained down from the gods and heavens above. Consequently, they believed that bees were the messengers and incarnations of the gods, who had bestowed honey from on high. A translation of one papyrus reads, "When Ra [a powerful god] weeps again the water which flows from his eyes upon the ground becomes a bee. They work in flowers and trees of every kind and wax and honey come into being from Ra's tears." Throughout the ancient kingdoms of Egypt, hieroglyphs of bees were used to signal omniscience, power, and deity.

Sacred writings of ancient India also associate bees and honey with the deities. *Madhu* means "honey" in Sanskrit, and the word is found frequently in the *Rig-Veda*, the sacred Hindu scriptures, which date from about 1500 B.C. The Vedas explain that the gods Vishnu, Krishna, and Indra were known to be "Madhava," meaning "honey-born ones," and that their symbol and incarnation is a bee. Lines of the Vedas describe Madhu coming from the clouds and the belief that bees were sweet liaisons between heaven and Earth.

The stories from ancient Greece are varied, but it was generally believed that bees were created by the gods of Mount Olympus. In the most common legend about their birth, the powerful god Zeus was in love with a beautiful girl named Melissa and turned her into a bee so that she could serve as nurse and consort to the gods for eternity. Melissa nurtured Zeus with milk and honey, the food of the heavens. Subsequent bees, called Melissae, were thought to live in the clouds and descend to Earth as caretakers, confidantes, and coconspirators of the gods. Working on behalf of the deities, bees were said to lead worthy pilgrims to the Oracle at Delphi. Honey, or *melis*, also came from the clouds, raining down into outstretched flowers before it was gathered and brokered by the Melissae.

The epics of Homer from the eighth century B.C. are liberally sweetened with references to sacred bees and their blessed manna.

In the *Odyssey*, the hero declares, "Tarry till I bring thee honey-sweet wine, that thou mayest pour libation to Zeus and all the immortals first." Honey alone and wine sweetened with it were used as tribute and offering throughout the ancient world. The *Iliad* describes bees' and honey's divine nature, and makes clear that admirers had already spent a good deal of time observing the creatures at work on Earth. "They swarmed like bees that sally from some hollow cave and flit in countless throng among the spring flowers, bunched in knots and clusters."

Aristotle was the first writer to attempt a more scientific, less mythical approach to the cult of bee admiration. Sitting next to a hive almost twenty-five centuries ago, he began the tradition of celebrating, observing, and exploring the secrets of honey and the hive that Smiley inherited. In the olive groves above Athens, Aristotle studied wild and domesticated bees, which at the time were kept in straw baskets, and recorded his observations in *Historia Animalium* and *De Generatione Animalium*. He described the bee's body as having "three parts, more than four legs, also teeth, a trunk, membranous wings and an interior sting at the rear." He was dead wrong about the teeth and close on the number of legs, but otherwise accurate on trunk, wings, and sting. Performing some of beekeeping's earliest experiments, Aristotle learned that "Neither wing nor sting will grow again if removed." And, "If one removes the heads of the grubs before they get their wings, the bees eat them up; if one nips off a drone's wings and then lets it go, the bees eat off the wings of the remaining drones." Aristotle attempted to record only what he had seen or discovered himself, but occasionally he was seduced by mythology and conjecture. On the origins of honey, for example, he writes, "The honey is what falls from the air, especially at the risings of the stars, and when the rainbow descends.... The bee makes the comb, then from flowers, as has been said, but honey it does not make; it fetches in what falls from the air." In his natural histories, Aristotle offered many such sincerely poetic inaccuracies, along with valuable facts and observations on the bee that are respected to this day.

Three hundred years later, Virgil joined the bee fan club. The *Georgics* picks up recorded apiculture where Aristotle left off. In the bee-filled lemon groves behind his house outside of Rome, Virgil read Aristotle's teachings and added useful details about hive construction, placement, and swarming. Virgil was less accurate in his reporting than Aristotle, more content to deliver a poetic fusion of fact, fiction, and myth. For the many things he couldn't observe or understand, such as where the honey or the bees came from, he confidently invented "facts" or blithely repeated what had become common knowledge. On the origins of honey, for example, he stated simply that it was "heaven-borne, the gift of air."

On the origins of the bee, Virgil presented the standard belief of the day, which was that baby bees were harvested from the flowers upon which the adults fed. "They indulge not in conjugal embraces, nor idly unnerve their bodies in love, or bring forth young with travail, but of themselves gather their children in their mouths from leaves and sweet herbs." Virgil relays another belief, also common, that bees were generated from the rotting body of a bull or lion. As strange as it sounds, this was not a new idea but a creation story accepted since ancient times. In the Old Testament of the Bible, for example, Samson "turned aside to see the carcase of the lion; and behold, there was a swarm of bees and honey in the carcase. He scooped the honey out with his hands and ate it as he went along."

This carcass-borne belief was still going strong in the tenth century, when it was recounted by a Byzantine writer named Florentinus. Although there are dozens of accounts, his is the most colorful and comes complete with instructions for man-made, rather than god-bestowed or flower-produced bees.

> *The method is this: Let there be a building 10 cubits high, and the same number of cubits in breadth, and of equal dimensions at all sides, and let there be one entrance, and four windows made in it, one window in each wall. Then bring into this building a bullock two and a half years old, fleshy, and very fat. Set to work a number of young men and let them powerfully beat it, and by beating let them kill it with their bludgeons, pervading the bones along with the flesh. But let them take care that they do not make the beast bloody (for the bee is not produced from blood), not falling on it with so much violence with the first blows. And let all the apertures be stopped with clean white cloths dipped in pitch, as the eyes and the mouth, and such as are formed by nature for necessary evacuation. Then having scattered a good quantity of thyme, and having laid the bullock on it, let them immediately go out of the house, and let them cover the door and windows with strong clay, that there may be no entrance or vent to the air nor to the wind.*
>
> *... Having then opened it on the eleventh day after this period, you will find it full of bees crowded in clusters on each other, and the horns and the bones and the hair and nothing else of the bullock left.*

Today, we know that bees absolutely abhor carnage and foul odors. It is unlikely that they would have been attracted to this method. Certainly they were not produced by it. More likely, centuries of dedicated experiment with bludgeoned bulls had produced blackflies or carnivorous wasps, which roughly resembled the dark Egyptian bee. Yet this recipe for bees is striking in the commitment, effort, and delusion it entails. The desire to understand and master the mythical creature was so great that men eagerly persisted in trying to fashion bees from flies and rotting flesh well into the nineteenth century. As late as 1842, a Mr. Carew claimed to have successfully performed the experiment in Cornwall, England.

In *The Lore of the Honey Bee*, Ticknor Edwardes describes the works of the ancient bee masters as an "ingenious leavening of a great mass of quite obvious fable by a very small modicum of enduring fact." This was the case from Virgil's antiquity throughout much of the Middle Ages. In the first century, Pliny the Elder devoted many pages of his *Natural History* to what was then known or imagined of bees, beekeeping, and honey. The latter, he thought, might be the "saliva of the stars" gathered by the bees. "At early dawn the leaves of trees are found bedewed with honey, and any persons who have been out under morning sky feel their clothes smeared with damp and hair stuck together, whether this is the perspiration of the sky or a sort of saliva of the stars or the moisture of the air purging itself... it brings with it the great pleasure of its heavenly nature."

Pliny's descriptions of the life of the hive are a delightful mélange of observed fact, compiled legend, and fanciful extrapolation. He writes, fairly accurately, that "they go out to their works and their labors, and not a single day is lost in idleness when the weather grants permission. First, they build combs and mould wax; in this way they build their new homes and cells." Next, he marries a morsel of truth to a whimsical invention: "They work within a range of sixty paces, and subsequently when the flowers in the vicinity have been used up they send scouts to further pastures. If overtaken by nightfall on an expedition they camp out, reclining on their backs to protect their wings from dew." Bees don't generally camp out on their backs at night unless they are dead, but Pliny—a famously busy man (who is said to have had a servant reading to him while he was carried about town and another recording his voluminous observations and opinions)—does not seem to have lingered until morning to find this out. He was charmed by the idea of bees as efficient little soldiers, draining honey from the stars and air. At nightfall, he suggested, bees were hushed by one bee "giving the order to take repose with the same loud buzz with which they were woken, and this in the matter of a military camp; thereupon they all suddenly become quiet." He adds more fantasy a few lines on: "Working bees catch favorable breezes. If a storm blows up, they balance themselves with the weight of a little pebble gripped by their feet. Some state that the stone is placed on the bees' shoulders." Pliny's understandings and fancies about bees (which borrowed heavily from Aristotle, Virgil, and other early masters) were standard for more than a thousand years; enthusiasts simply accepted and repeated what they heard about these intriguing creatures, often without ever going near a beehive.

Starting in the middle of the fifteenth century, fable and fancy gave way more and more rapidly to fact. Truth and progress in beekeeping were aided by the invention of the printing press in the 1450s. The works of Aristotle, Virgil, and Pliny the Elder were immediately printed, as was Columella's *De re Rustica*, a compendium of agricultural and beekeeping knowledge. Although slow by today's Internet standards, the printing press allowed information, enthusiasm, and exchange to spread like Renaissance wildfire, and the bee's fan club flourished. All over the beekeeping universe, men (and a very few women) observed, experimented, and swapped ideas as they explored the secrets of the hive. In the farmyard, where the valuable household honey supply was obtained, and in the halls of academe, where bees yielded intriguing scientific clues, the honeybee was of immense interest and subjected to intense scrutiny. In England alone, dozens of substantial books about beekeeping were published in the sixteenth and seventeenth centuries. Beekeeping innovations and inventions multiplied competitively as enthusiasts, academics, and businessmen flocked to the new science and frontier of bee husbandry. The British Patent Office opened in the year 1617, and the 108th grant in the land, in 1675, dealt with managing bees and their produce. At about this time, King Charles II appointed the very first Royal Bee Master, Moses Rusden. In 1679 Rusden published *A Further Discovery of Bees*, which was, for the seventeenth century, a best seller.

By the middle of the seventeenth century, through obsession, observation, experimentation, and aggressive publishing, it was generally acknowledged that the hive had a ruler, presumably a king, who differed in size and behavior from the other bees. The architecture of the hive, and the work schedules and seasonal cycles of the various inhabitants, were roughly understood. The inner workings of a living colony—life spans, biological and reproductive functions, and detailed anatomies of the residents—remained tauntingly mysterious.

Until this time, bee enthusiasts were forced to learn about the hive, more or less, from the entrance, like paparazzi or like morticians performing autopsies on dead, destroyed nests. Innovative snoops began to put walls of glass onto common hives, and by the end of the seventeenth century, peepholes into the life of the colony had been fashioned. Samuel Pepys wrote in his journal in 1665 that "after dinner to Mr. Evelyn's; he being abroad we walked in his garden, and a lovely noble ground he hath indeed. And among other rarities, a hive of bees, so as being hived in glass, you may see the bees making their honey and combs mighty pleasantly."

Mr. Evelyn had most likely affixed a small sheet of glass or a bell jar to the top or side of a wooden box hive, allowing viewers a murky glimpse of a few outer combs of honey and some worker bees. With such innovations, enthusiasts progressed from the transom of the hive, as it were, to the foyer. The boudoir, the nursery, and other private regions of the domicile were still cloaked in mystery. Obviously, the biggest bee, the king, was thought to possess the keys that would unlock these domains, and the ruler was watched and tampered with obsessively.

Pepys's countryman, Charles Butler, got or imagined a glimpse into the royal nursery in 1609, when he asserted that he had seen the biggest bee laying eggs. At about the same time, the Spaniard Luis Méndez de Torres observed the same thing. Butler's and Mendez's claims about the sex of the queen seem to have been largely ignored until Jan Swammerdam, a Dutchman, used a microscope to verify them. As the only insect so sweetly exploited by man, and as popular and voguish as they were in the seventeenth century, bees went under the microscope soon after it was invented in 1608. Just sixty years later, Swammerdam dissected and inspected the biggest bee and its entourage under a powerful lens, presenting undeniable ovarian proof that the king was in fact a queen. In his *Historia Insectum Generalis*, he produced woodcuts and engravings that are to this day excellent illustrations of the anatomy of the three classes of inhabitants of the hive. A biography of Swammerdam from the time describes his dedication to research, a hallmark of those devoted to the bee. "This treatise on bees proved so fatiguing a performance, that Swammerdam never afterwards recovered even the appearance of his former health and vigor. He was most constantly engaged by day making observations, and as constantly by night in recording them by drawings and suitable explanations." This fatiguing performance cost the great Dutch explorer his health, but it yielded, at last, proof of the queen's sex. The question of who or what fertilized the eggs in the imperial ovaries, however, went unanswered by Swammerdam and his immediate successors.

Almost a century later, the renowned Swiss naturalist François Huber was still struggling to understand the royal methods of reproduction. Ironically, Huber, one of the great observers of bee culture, was blind. Born in Geneva in 1750, he lost his sight in adolescence from an illness. By adulthood, he had been seduced by bees, and he and his faithful (and seeing) servant François Burnens devoted their lives to studying them. With the help of family money, time, and teamwork, they made enormous contributions to the body of bee knowledge. Huber continually suggested experiments, and Burnens tirelessly carried them out.

One of this team's great contributions was the development of a hive that allowed more access to the inner workings of the colony than ever before. Many different configurations for study had been attempted by scientists all over the world, but the Huber duo developed a revelatory observation hive, in which internal frames were hinged together on one long side as if they were leaves in a book. Experienced beekeepers could browse through a colony by prying apart and turning the honeycombed pages. In his journal, Huber (or Burnens) wrote of the leaf hive: "Opening the different divisions one by one, we daily inspected both surfaces of every comb; there was not a single cell where we could not see distinctly whatever passed at all times, nor a single bee, I would say, with which we were not particularly acquainted."

Observation hives allowed scientists and bee voyeurs greater access to the inner workings of the hive. They were also a source of entertainment for an avid, admiring public. An engraving from a French beekeeping manual published in 1740 shows elegantly dressed ladies and gentleman standing in an outdoor apiary,

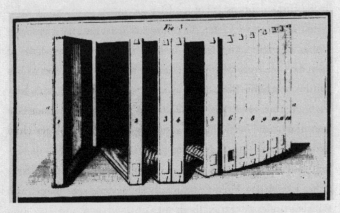

Drawing of the Huber leaf hive, from Huber's 1821 book,
New Observations on the Natural History of Bees. The illustration
shows the hive open like a book for "reading" the frames or leaves within
(which would be filled with honeycomb and bees).

A 1740 book shows a lady and gentleman peering into the observation windows of some hives, while a beekeeper captures a treed swarm in an old-fashioned straw hive, called a skep.

peering intently into the eye-level windows of tall hive structures as if they were periscopes onto another world. Eva Crane, in *The World History of Beekeeping and Honey Hunting*, describes an exhibition in England in 1812 that offered "six beautiful glass beehives, with their appendages, containing a complete swarm in each hive ... so arranged that the spectator can at one view fully comprehend the truly wonderful order, contrivance and harmony that pervades this astonishing community." Admission cost a shilling each for ladies and gentlemen, and sixpence per child and servant.

An illustration of a Ukrainian observation hive from the 1840s shows an enraptured family gathered around a glass-walled box flush with bees and comb. From the 1850s in England comes an illustration of a bell-shaped jar full of bees placed on a fancy drawing room table. A tube running to the window appears to be connected to a beehive located just outside.

Observation hives are still used for study and entertainment. In the modern version, a frame, preferably the one on which the queen is holding court, is taken temporarily (for show-and-tell period only) out of the hive and placed in a slim wooden carrying cabinet between two panes of glass. It's a slice of hive in a clear briefcase, an ant farm with bees, and a perfect way to see them busily at work. Beekeepers often take them to fairs and festivals to attract, educate, and sell more honey to their enchanted customers. Smiley's beekeeping stories and his tupelo are usually enough to enthrall his clientele, but sometimes he borrows an observation hive for big honey-selling opportunities or when he goes to speak at agricultural fairs. "Kids just love that thing," he says. "Everybody wants to know about the bees."

Thomas Wildman expanded the bee's celebrity (and his own) by strolling around London wearing a beard of bees and inducing his swarms to perform tricks, which delighted the gathered crowds. In 1748 he wrote *A Treatise on the Management of Bees*, in which he reluctantly revealed his secrets.

My attachment to the queen, and my tender regard for her precious life, makes me most ardently wish that I might here close the detail of this operation, which, I am afraid, when attempted by unskillful hands, will cost many of their lives; but my love of truth forces me to declare, that by practice I am arrived at so much dexterity in management of her, that I can, without hurt to her, tie a thread of silk round her body, and thus confine her to any part in which she might not naturally wish to remain.

In order to walk an entire swarm through town, as Wildman did, he enlisted the cooperation and pheromones of the queen. For a wig or beard made of bees, modern wranglers do the same, placing the queen in a screened box and positioning it on their pate or neck. Experienced beard growers apply sugar water to their chins or bathing-capped heads, plug their nose and ears with cotton, dab some insect repellent around their eyes, then quietly wait for the bees to do what comes naturally, which is flock to their queen. They are content to be near her and crawl gently around, licking up the sticky sugar water, thrilled not to have to travel and work for it. The longest bee beard that I have heard of was over three feet long and in place for twelve minutes. I have not tried this trick yet, but I suspect it takes some patience and tickles quite a bit. The *Guinness Book of World Records* reveals another (and even stranger) bee-related feat: in 1998, a man kept 109 honeybees in his mouth for ten seconds to achieve the world record for "the most bees in the mouth." This I do not want to try.

Although the queen had become a public spectacle, Huber and Burnens remained fixated on her personal life. Through endless, sometimes ruthless experiment, they got to know her more intimately than had any previous admirers. The team's many discoveries were revealed in Huber's *New Observations on the Natural History of Bees*, a series of scholarly letters that was published in Geneva between 1792 and 1814. In the opening missive, he reveals his challenge, understanding the queen bee's reproduction: "Having now come to the particular object of this letter, the fecundation of the queen bee, I will describe the new experiments by which I think I have solved the problem." He then describes one of the hundreds of studies that he and Burnens were conducting in their bee laboratory outside of Geneva:

At eleven in the forenoon, we placed ourselves opposite to a hive
containing an unimpregnated queen five days old.... The males
began to leave the hives. Soon after, the young queen came to the
entrance; at first she did not fly, but during a little time traversed
the board, brushing her belly with her hind legs.... She then flew
away, describing horizontal circles twelve or fifteen feet above the
earth....we placed ourselves in the center of the circles described
in her flight, but she did not remain long in a situation favourable
for our observations, and rose out of sight.

The queen returned in exactly twenty-seven minutes. "We now
found her in a state very different...the organs distended by a substance, thick and hard, very much resembling the matter in the
vessels of the male; completely similar to it indeed in colour and
consistence."

After several repetitions of this experiment, Huber concluded
that the queen was "fecundated" with sperm from the male drones.
Furthermore, he had proof that this mating event happened outside the hive, just once or twice, before the queen retired indoors for
the rest of her life. But how could she reproduce for a lifetime after
copulating just that once? How was it that she laid mostly female
worker eggs, alternated with a regular percentage of male drones?
Huber and Burnens were fascinated, obsessed even, but no matter
how many dizzying mating flights they watched, they remained
frustrated about many of these intimate royal details.

All over the world, ardent admirers were prying into the queen's
private life. As Huber's research was ending in France, a German
Catholic priest named Johannes Dzierzon began investigating the
monarch's fertility. He too observed the mating flight and, upon

dissection of the "fecundated" queen, discovered a small sac filled with sperm, which he called the *spermatheca*. This bulging reservoir of the "male substance" explained how the queen could mate only once and produce thousands of eggs a day for several years. The royal sperm bank was conveniently located on the busy highway of the queen's oviduct, so fertilizer could be dropped onto eggs as they slid by.

Dzierzon next turned his attention to the male substance. Developing experiments to "retard" the impregnation of the queen, he made significant discoveries about the use and effect of the sperm. Dzerzion put several queen bees on ice after their mating flight and found that, with the sperm thus incapacitated, they subsequently produced only male bees. A colleague, Berlepsch, tested this theory in a similar experiment. He "refrigerated three queens by placing them thirty six hours in an ice-house, two of which never revived, and the third laid, as before, thousands of eggs, but from all of them only males were evolved. A short exposure of a queen, to pounded ice and salt, answers every purpose. The spermatozoids are in some way rendered inoperative by severe cold." From this type of chilling experiment, Dzierzon and Berlepsch determined that a spermless or dysfunctional queen would produce only male eggs. A properly fertilized monarch, on the other hand, had a choice. She could lavish sperm onto eggs to produce female workers; she could also withhold it to produce the few hundred males necessary to the colony.

For every renowned Johannes Dzierzon and Amos Root, countless enthusiasts around the world were conducting their own backyard experiments. Anyone with a swarm of bees could participate. By the middle of the nineteenth century, after nearly two thousand years of devotion by scholars and backyard beekeepers alike, many of the most pressing questions of the hive had been answered. The basic anatomy and most of the nuanced functions of each member of the colony were understood. Beekeepers were aware of how the queen reproduced, and of the life cycles and foraging habits of her offspring.

As beekeepers mastered the habits and biology of the hive, their inquiry turned more and more to extracting the precious produce. Since Aristotle, people had had difficulty entering a colony to observe, husband, and harvest without damaging the hive and bees. In a constant quest for structural strength, bees will seal any unused space in the hive with wax and propolis, their sticky household glue made from plant resin. This efficiency essentially solders all of the interior parts of the nest to the walls, making the bees' home stronger and its food stores more inaccessible. In most hives (natural or man-made) in most centuries, the comb had to be messily cut out to be inspected or harvested, a violence that disrupted and often destroyed the colony. Around the world, beekeepers yearned for a way to steal the honey and spare the hive.

Lorenzo Langstroth, one of the great bee thinkers and tinkers, devoted himself to finding a solution to the great honey problem. As a tutor at Yale, and then as a minister of the South Church in Andover, Massachusetts, he also devoted himself to religion. Historically, houses of the spirit also have been incubators of beemasters. Ancient temples and monasteries typically had apiaries, and both bees and gods seem to attract types devoted to quiet observation, contemplative thought, and a tireless search for answers. During his ministry, Langstroth suffered from what were vaguely called "head troubles," accompanied by bouts of hysterical muteness that forced him eventually to resign. He moved away from his wife and son and turned his attention to teaching school and studying bees. Like many bee masters, he was eccentric and indefatigable in trying to fathom the secrets of the hive and wrest more honey from it. From his troubled head came one of the culminating discoveries of beekeeping and honey harvesting—bee space.

For many years, various experts had observed that bees were very particular about their space. If there was a small gap in the hive, the bees immediately and efficiently sealed it with propolis. A larger opening would inevitably be filled with wax comb. In his hives (for a time he had Huber leaf models), Langstroth observed that the bees left a very exact gap of three eighths of an inch—or the comfortable width of a worker—open between combs for hive traffic and maneuvering. He was not the first student to notice this gap, but he was the first to realize how it could be applied to the beekeeper's and harvester's advantage. In his barn in Oxford, Ohio, he devised a hive in which this delicate margin of bee space existed between each frame, and also between the frames and the walls of the hive. Describing the moment of discovery, which occurred on the day of October 31, 1851, he wrote:

> *Pondering as I had so often done before, how I could get rid of the disagreeable necessity of cutting the attachments of the combs from the walls of the hives . . . the almost self-evident idea of using the same bee-space as in the shallow chamber came into my mind, and in a moment the suspended movable frames, kept at a suitable distance from each other and the case containing them, came into being. Seeing by intuition, as it were the end from the beginning, I could scarcely refrain from shouting out my 'Eureka!' in the open streets.*

The Reverend John Thorley, preacher and bee enthusiast, at his desk
composing *Melissalogia*, a study of bees. His subject can
be seen on the desk in front of him and in skep hives outside the door.

Langstroth's epiphany meant that the bees did not solder the
combs to the hive interior with glue or wax. It sounds simple now,
like Gutenberg's movable type, but it changed the world of honey.
Full frames could be simply lifted out of the hive and inspected,
emptied, or manipulated without the making of irreversible, wound-
ing cuts. The beekeeper no longer had to destroy the hive to observe
his livestock or harvest his honey. Langstroth's book, *Langstroth on*

the Hive and the Honey Bee: A Beekeeper's Manual, published in 1853, was a seminal guide for the first generations of bee-space-savvy beekeepers.

The Langstroth hive put the beekeeper finally and completely in charge of his livelihood, able to manipulate and harvest at will, with minimal mess and death. Langstroth wrote of his invention: "The chief peculiarity in my hive was the facility with which [the frames] could be removed without enraging the bees.... If I suspected anything was wrong with a hive, I could quickly ascertain its true condition, and apply the proper remedies.... The use of these frames will, I am persuaded, give a new impetus to the easy and profitable management of bees." He was right. Professor A. J. Cook, the author of a Langstroth-inspired beekeeping manual that came out twenty-five years later, wrote: "It is this hive, the greatest apiarian invention ever made, that has placed American apiculture in advance of all other countries." With a few intervening refinements, this is the type of hive used today by Don Smiley and most modern beekeepers around the world. My own amiable bees and plentiful harvest are the result of Lorenzo Langstroth's innovations and head troubles.

In the beginning of the twentieth century, a Belgian dramatist named Maurice Maeterlinck studied the hive and was rewarded with poetry. He wrote *The Life of the Bee* after years of observing bees as he lived amongst them. His devotion took place in France, surrounded as he was by orchards and vineyards and as he sat at a table upon which, like Aristotle, he had placed little dishes of honey to attract his muse. Using an observation hive and pots of paint, he marked individual bees with bright dots of color and chronicled the life of the colony in lyrical, spiritual detail. Maeterlinck did not contribute scientific advances, but his little book, only 150 pages long, is a marvel of natural history writing, courtesy of the bees. His introduction describes the first apiary he saw and how he fell in love with the bees and became a student of the hive.

> *Here, as in all places, the hives lent a new meaning to the flowers and the silence, the balm of the air and the rays of the sun. One seemed to have drawn very near to the festival spirit of nature. One was content to rest at this radiant crossroad, where the aerial ways converge and divide, that the busy and tuneful bearers of all country perfumes unceasingly travel from dawn unto dusk.... One came hither, to the school of the bees, to be taught the preoccupations of all-powerful nature, the harmonious concord of the three kingdoms, the indefatigable organization of life, and the lesson of ardent and disinterested work; and another lesson too, with a moral as good, that the heroic workers taught there and emphasized, as it were, with the fiery darts of myriad wings, was to appreciate the somewhat vague savor of leisure, to enjoy the almost unspeakable delights of those immaculate days that revolved on themselves in the fields of space, forming merely a transparent globe, as void of memory as the happiness without alloy.*

The Life of the Bee was a best seller when it was published in 1901, becoming the most popular book ever written about bees (or any type of insect; perhaps only the novel *The Secret Life of Bees*, published in 2002, has supplanted this). Ten years later, Maeterlinck won a Nobel Prize for literature, encouraging book sales and setting off another wave of enthusiasm for the honeybee. Karl von

Frisch caught the rapture in the 1920s, when he started studying bees at the University of Munich. Fifty years later, he won a Nobel Prize for his research in another frontier of study: the language of bees. Twentieth-century science was of course helpful to von Frisch, but really his methods and motivation were not unlike Aristotle's, Huber's, or Smiley's—patient hours of watching, wondering, and admiring. In his Nobel Lecture, he describes all three:

> *In order that the behavior of foragers could be seen after their return to the hive, a small colony was placed in an observation hive with glass windows, and a feeding bowl was placed next to it. The individual foragers were marked with colored dots, that is, numbered according to a certain system. Now an astonishing picture could be seen in the observation hive: even before the returning bees turned over the contents of their honey sack to other bees, they ran over the comb in circles, alternately to the right and the left. This round dance caused numbered bees moving behind them to undertake a new excursion to the feeding place.*

He goes on to explain the specifics of the dances that bees used to communicate.

> *The tail-wagging dance not only indicates distance but also gives the direction to the goal. In the observation hive, the bees that come from the same feeding place make their tail-wagging runs in the same direction, whereas these runs are oriented differently from bees coming from other directions. However, the direction of the tail-wagging runs of bees coming from one feeding place does not remain constant. As the day advances the direction changes by*

the same angle as that traversed by the sun in the meantime, but in
the opposite rotation. Thus, the recruiting dancer shows the other
bees the direction to the goal in relation to the position of the sun.

And what if there were no visible sun? Von Frisch determined that, because of their sensitivity to ultraviolet light, bees can detect the position of the sun even on the cloudiest of days. Conversely, near the equator, when the bright orb is directly overhead, bees are unable to dance and must wait until the sun is degrees away from its zenith before they can resume shimmying in its shadows. Von Frisch cataloged a complex, precise vocabulary of the dances the bees used to convey the location and quality of food sources. A circular dance, for example, indicates food close to the hive, while a figure-eight dance signals resources up to a mile away. The duration, angle, speed, and configuration of their jig can also broadcast fear, alarm, or happiness. Von Frisch's observation, experimentation, and curiosity answered yet another mystery of the hive, ten thousand years after men had asked the first questions.

Part of an ancient brotherhood that includes Aristotle and von Frisch, Donald Smiley is constantly watching and learning from his bees and devising ways to improve his husbandry and harvest. "There's not a frame in any hive that hasn't been in my hand at least once this year," he says. "That's how you manage bees, you get to know them. You get to know what's going on in the life of the colony. I have a pretty good idea, but there's always more to learn. Not a year goes by that I don't see something different, learn something different. I never get tired of this." Reading the writings of Aristotle and Huber, one has the impression they would have said the same thing.

In the years since that first painful harvest, Smiley has learned his bees and his business well. One of the local experts, he even makes house calls on occasion, visiting yards and helping beginners with their hives and equipment, sharing his knowledge and enthusiasm. Unlike the first debacle, he can now walk into a beeyard and know exactly and instantly what to do. He takes a deep breath, listens quietly, watches a flight pattern or two, and calmly goes to work. "You know what I'd rather be doing now?" Smiley asks with a grin as he looks at a healthy frame dripping with bees. "What I'm doing now."

FEAR

To obtain my own livestock, I called the dealer who had supplied my friend Ace, and was told to come collect them the first weekend in May, which is typically when trucked shipments of live bees arrive in the Northeast from their nurseries in the South. When I pulled into the dealer's driveway, where a pyramid of boxed bees loomed in the blue shade of a tarpaulin, I was suddenly full of trepidation and questions. Reading about bees at the library is very different from taking several thousand home to live with you. These were real. These were crawling, stinging, flying, buzzing, fierce-looking bees. Politely and professionally ignoring my visibly shaken nerves, the dealer plunked a package of Italian bees into the back of my car. The screened-in wooden box, about the size of a cinder block or a large shoebox, housed a dark brown, dense, writhing mass of about ten thousand insects vibrating and humming in a way that seemed ominous. Somewhere in the center of this daunting cluster was a queen, caged for her safety during transport and for my supposed ease of handling. A few strays orbited the box and boldly began checking out the interior of my car, at which point I seriously considered having a more hands-off, academic love affair with bees.

To discourage any of those orbiting scouts from coming up to visit me, I drove home too fast with all the windows down, blaring the radio to drown out their frightening whine. For once, I wanted to get stopped by the police so that when they asked why I was speeding I could simply point to the teeming box in back. Arriving home, I left the beast waiting in the shady back of the car while I stalled by making sure my apiary was correctly set up. The bees' future establishment was about eighty feet from my house on a grassy hillside, obscured from view by the crest of the hill but visible from my upstairs bedroom window in case I wanted to spy on them from there. I had painted the box a buttery white to deflect the heat of the sun, and I had chosen a spot where the bees would get plenty of sun, water, and protection from the northern winds, just as my manual and the ancients had suggested.

My stalling complete, I reread the instructions in my bee manual, Ed Weiss's *The Queen and I*, and sprayed the bees repeatedly and overzealously with sticky sugar water, which is supposed to cool and calm them and make it difficult for them to fly with awkwardly sodden wings. Spraying had the effect of immediately turning down their volume, and I noticed that they became preoccupied with licking the sugar off each other in a languorous group snack. I sprayed them again. Weeks before the pickup, in a burst of macho fearlessness and bee love, I had decided I didn't need a protective white suit. Now I was regretfully bundled up in jeans, boots, and a long-sleeved shirt with my new canvas-and-leather beekeeping gloves pulled tight to the elbow. The veil of my shiny white plastic beekeeper's hat was tucked into the buttoned-up collar of my cotton shirt, the tails of which were pushed down into my jeans as far as they could go. I was an underdressed rookie alone with thou-

sands of bees who had to be relocated somehow from their traveling case to their new home. Despite the drugged indifference of
the bees and the newness of my veil, I was scared, aware of every
hole and wrinkle in my armor.

By the time I got back to the hive, carrying the vibrating box of
bees, the only thing I could hear was my pulse echoing in my plastic
hat. Though I had memorized the instructions, I clutched the sticky
manual and rehearsed the plan aloud one more time. Remove caged
queen and traveling food supply. Place caged queen in hive. Pour
her thousands of subjects in after her. No problem. Taking a deep
breath, I pounded the box gently down on the ground, and the bees
settled with a roar to the bottom of the box, sedated with sugar
water and heat.

Fumbling in my stiff new gloves, which now seemed as grace-
ful as oven mitts, I slid the lid on the box to the side just long
enough to extract the food can and the queen cage suspended
within. Encouraging myself loudly, I discarded the can on the
ground. Pushing two brads into each top side of the little cage, I
hung the queen down between two frames in the deep hive body.
I felt (and looked) as though I were in New Mexico handling ura-
nium, yet the swarm of bees in the screened box seemed to understand the precious nature of the caged cargo. The shoebox of bees
was queenless now and agitated for real, no matter how swamped
in sugar water. They had met their ruler only a couple of days earlier, when a dealer in Georgia had inserted her caged entourage
into their midst, but already they were devoted, in thrall to her
pheromones.

As the queen awaited her clamoring subjects, I considered call-
ing Ace to help me with the next part, in which I was supposed

to upend the box and pour the mass of sticky bees into the hive through a round hole as wide as a soup can. Repeating my mantra of instructions and the assertion that bees don't sting unless provoked, I looked down at the box of provoked, oversprayed, queenless bees and started considering anaphylactic shock for the first time. I wondered if I would have enough time or function after a sting to drive myself to the hospital.

In one breathlessly clumsy maneuver, I rolled the box over and shook it gently over the hive. The bees plopped and oozed down onto the frames like harmless gravelly mud. They spread over the frames and slipped into the box like mercury, paying no attention to me whatsoever, interested only in reuniting with their queen and getting started on the business of the hive. The first thing they would do was free the monarch from her cage, which was plugged with a tasty paste of confectioners' sugar and water. By the time they licked her out, in a day or two, her pheromonal spell would be firmly cast, her powers absolute. They seemed to go right to work, licking and liberating. I began to breathe again. There was still a dazed, sodden clump of bees on top of the frames, and gathering my courage, I reached out to smooth and stroke it as if it were a mound of earth in a garden. The clump responded as amicably and docilely as a new puppy and rolled down into its new home, taking all of my fear and anxiety with it.

The installation was complete: a stingless, harmonious piece of cake. With reluctance, I lowered the hive cover into place, following instructions to let the bees recover and get acclimated and acquainted for a week before another meddling disruption. I knew it would be hard to wait that long to visit again with my new guests. As if coasting to the end of a thrilling roller coaster ride, I immediately wanted to go again.

Later that evening, I went and sat next to the hive as the sun was setting and cool breezes were arriving. Bullfrogs began their nightly chorus of barks, and the weeping willow behind me whispered as I watched several bees at the entrance. They had shaken off the trauma of the day and the journey from Georgia, and had already been out investigating their new digs, eager to get started finding food for the colony. They seemed to be humming as they worked, and their pleasing alto murmur entwined with the basso of the bullfrogs, the falsetto of the tree chirpers, and the rustling percussion of the willows. This jazzy ensemble performance was the sound track for my dreams that night.

The next morning, coffee cup in hand, I sleepily sat next to the hive and saw that the bees were already active. As I watched, several returned to the landing area with bright yellow pollen gathered on their rear legs. Entering the hive, they passed sisters intently darting out and off, perhaps going to investigate the dandelions studding the hillside or the fragrant lilies of the valley clustered next to my house. I wondered if all the local treasures had already been discovered, and gossiped and danced about in the hive, for the residents seemed intimately and purposefully acquainted with every tree and flower in the landscape. I could see them plunging into plants everywhere I looked. On the hive stoop, a few bees lingered, cooling themselves before entering their busily hot domain. Next to the entrance, I noticed one bee lounging quietly on the tip of a tall blade of grass, bending it slightly. She seemed to be taking the morning off in her bee hammock, but most of her kin went determinedly about their work as if they had lived in this patch of Connecticut their entire lives. They blithely ignored me. I could have been a pebble, a piece of plastic, or some other nectarless object. It was hard to believe I had ever been afraid of them. It was hard to believe that my house and garden had ever felt complete before the bees arrived.

THREE

Robbing the Bees

Now there was honey comb in the countryside; Jonathan stretched out the stick that was in his hand, dipped the end of it in the honeycomb, put it in his mouth, and was refreshed.

I Samuel 14:25–7

If you want to harvest honey, don't kick over the beehive.

Abraham Lincoln

The only reason for being a bee that I know of is making honey... and the only reason for making honey is so I can eat it.

A. A. Milne, Winnie-the-Pooh

By the beginning of May, the tupelo buds have exploded into blossom, and Smiley's bees are determinedly at work storing tupelo nectar. Smiley is now on an unknown but daunting deadline, obsessed with harvesting maximum tupelo before the flowers fade away. As if chasing a downpour with an organized collection of buckets, he must have all 700 hives in the right places at the right time and be able to empty and refill them during the deluge without missing too many precious drops.

This annual juggling act requires some assistance, which arrives tousle-haired every morning in a battered green pickup truck in the person of George Wolinski, a twenty-year-old Wewahitchka native who has been working in bees, on and off, since he was fourteen years old. Like many of the men in the area who work outdoors, George is scrawnily thin and deeply tanned. He is dark-haired, with a shadowy mustache struggling to add age to his boyish face, and reports for duty in a work uniform of tattered blue jeans, faded work shirt, and a baseball cap. He usually brings his own lunch or snack, which consists of a can of soda and a pack of cigarettes. His grandmother is Smiley's much older sister, which they believe makes George his great-nephew. Smiley just calls him his nephew, or Georgie, or "son" (the way he addresses most men younger than himself). Georgie is possessed of a surprising strength, an engaging smile, and unflappable calm. "He's not afraid to work, and he don't back away from nothing," says his boss. These are two much-needed and elusive qualities in beekeeping help. Even though Smiley pays a good hourly wage, workers willing to spend days in the sweltering sun lifting heavy boxes while enduring repeated stings are hard to come by. He tells many anecdotes of help gone AWOL and awry. "Remember that one guy who said his knees couldn't take it?" asks Smiley, grinning. "Remember how he just walked off the job in the middle of it and then walked six miles home on those bad knees?" He and Georgie both smirk and shake their heads. The joys and challenges of beekeeping aren't obvious to everyone, though they seem to have captivated George. "I had a job in Panama

City installing glass before I started doing this," he says. "I reckon I learned just about everything there was to know about glass. But here, I'm still learning. I like what I'm doing. I may even get my own hives some day." For the usually quiet, taciturn George, this constitutes a wordy, passionate manifesto.

Today, George has recruited his friend Keith to help with the harvest. Having assisted his grandfather with his bees a few times, Keith is slightly experienced, but when they get to the yard, he looks askance at the large and possibly painful scale of Smiley's operation. The trio is standing in a small clearing tucked at the end of the East River Road, which begins off Route 71 near the high school. Under the morning shade of three enormous live oak trees and a few tall pines are sixty-four hives stationed for the tupelo flow. Two sides of the yard are hemmed by bushes heavy with pink, unripe blueberries. On another is a decaying white house with a dilapidated gray pickup truck parked in front. The fourth border of the yard is water. The Chipola, a tributary of the Apalachicola, chugs thirty feet away, at the base of a steep, crumbling embankment. Across the swirling chocolate brown river lies a dense, low island of tupelo trees. Bees leave the yard and dart purposefully out over the water, headed for the bountiful nectar buffet on the other side.

The buzzing in the yard has a noticeably high pitch to it, and the bees seem to be flying faster and straighter, trumpeting their excitement over the annual feast. Smiley does a quick hive inspection and determines that these supers, stacked two and three high, are full of light, sweet tupelo honey collected in the last three or four days. "On a good tupelo flow, they'll fill a box and cap it overnight," he says admiringly. Every frame he lifts out is laden. He appraises the stacks in the yard and calculates about eight barrels, or more than two tons, of tupelo if all goes well. Pulling his hat and veil a little lower on his head like a cowboy, he says, "Let's rob some bees, boys."

His flatbed truck is loaded with sixty-four empty honey supers, dollies, blowers, and an assortment of gear. On the remaining surface of the truck bed, he lays out a dozen black wooden fume boards—hive lids lined in absorbent black felt—as if he were dealing cards. Onto each, he drizzles Bee Go from a plastic squirt bottle—a splash of butyric acid, which is normally found in rancid butter. Concentrated in Bee Go, the acid is more sour and intense than the most rotten dairy product any human has ever smelled, fouler even than the best smelly cheeses. It is benign except for its odor, which is vile enough to drive the bees quickly from their homes. Beekeepers with more time and fewer hives use bee escapes: one-way exits placed between the honey super and the hive body that block the bees from returning to their food stores on harvest day. Both escapes and butyric acid save beekeepers the time and nearly impossible task of having to smoke, brush, and pluck millions of reluctant bees from thousands of frames of honey. The hurry and timing of the tupelo do not allow Smiley the leisure of placing escapes and waiting for all the bees to be shut out of their supers. He wants this honey now.

Smiley directs his two-man crew to help him remove the hive covers and deal the smelly fume boards onto the tops of the first several colonies. The enveloping heat of the day accelerates the stench of the boards, and in a few minutes bees ooze and boil out of the bottom entrance as if a plunger had been pushed. Though eager to escape the noisome smell, the bees are not willing to abandon their homes completely; they spill out the door and crawl up the front of the hive box, enjoying the fresh air and the rays of the sun.

Almost as soon as the fumes are applied, the white hive faces are thick with writhing beards of bees. Some strands are long enough to trail to the ground, and Smiley steps over them carefully as he works the yard. "Don't anyone step on my bees," he warns.

As if he were painting the boxes, or shaving them, Smiley uses his bee brush, an eight-inch row of supersoft, two-inch-long yellow plastic bristles embedded in a wooden handle, to wipe the beards gently to the lower parts of the hive stacks. After he removes the

Donald Smiley robbing his bees.

upper boxes of honey, he'll leave these nest or brood layers of the colonies and their bees in the yard to gather more nectar. He'd prefer to leave the bees in their homes rather than take them to his, so he brushes them down, wafts a bit of smoke, and lifts each honey super on its end to ascertain that the bees have evacuated the top layers of the hive. "The trick is to leave as many bees in the yard as possible," he says, perhaps remembering his first harvest. With the exception of a very few stragglers, the Bee Go has done its job. After a nod to George, the workers whisk away the full supers, which weigh between seventy and ninety pounds, and transfer them to the truck bed.

For the few supers that still have too many bees clinging to the frames within, Smiley will use the bee blower stored in the back of the truck. Like a giant hair dryer on wheels, a blower has pretty much the same capabilities, except for heat settings. A hose attachment and nozzle directed between the frames catapult all the stubborn remaining bees out with hurricane-force winds, but because blowing is noisy, frightening, and disruptive to the bees, Smiley resorts to it as infrequently as possible.

As quickly as George and Keith load the full honey supers onto the back of the truck, they replace them with empty ones on the hives to catch the next liquid load of nectar. The yard is a rotating assembly line, with fume boards, workers, and blower leapfrogging down the line. Smiley orchestrates the precision teamwork that smokes, brushes, removes, and replaces the supers. No matter how efficient and organized the crew is, however, time and space in the yard are tight and tensions can run high in both man and insect. As in any robbery, the longer they stay, the more chance there is of trouble, and the more they distress the bees. Midmorning, as he strokes a beard of restless bees, he says gently, "We're messin' up y'all's day, ain't we, girls?"

By late morning the sun is blazing. The yard is hot and humid, thick with the smell of rancid butter, smoke, honey, and crushed grass and pine needles, and noisy with agitated bees, gentle crickets, and the grunts of the workers as they lift and move heavy supers around the yard. Patches of sweat blossom on their dark work shirts as the men perform their strenuous harvest dance. The pale edifice of supers on the back of the truck grows larger as the stacks in the yard are diminished from several towering stories to two, a deep hive layer topped by a honey super with ten empty frames.

In about two hours, all of the colonies in the yard have been robbed and resupered. One hundred and twelve boxes containing over a thousand frames full of honey are loaded and ready on the truck. George and Keith (who by this time is an old hand) drape a tarpaulin over the block of wood and honey to prevent the bees from stealing back their food while it remains in the yard. They tie the load down with ropes, hop into the truck, and trundle slowly back through town with the fragrant load. Orphaned bees orbit the freight as it journeys the three miles to the honey house. Eventually, these strays will give up on rescuing their honey and find their way back to their homes on East River Road, guided by their internal flight map and the alluring scent of their queen.

When the truck arrives back at Smiley's place with a mountain of supers, a few stray bees, and three sweaty humans, the honey house springs to life. Smiley and his two assistants fling open the wide screen doors, flip on the spinal strip of fluorescent lights, turn on the fan, and get to work. Stack by stack, the formation from the truck is dollied into the house and reassembled close to the uncapping machine, which looks like a giant food service toaster. George pulls frames from the boxes and lays them flat on the wire conveyer belt that inches toward the maw of the uncapper. Midway down the belt, underneath a stainless steel hood, are rods outfitted with chains three links long. When the rods spin, the sharp metal fringe shears the wax caps off the comb, top and bottom. As metal hits wood, wax, and honey, the uncapper whines like a table saw. On the loading end of the belt, the frames look like neat trays of cake with a glossy light icing of wax. Ten seconds later, after they have met the chains, they emerge looking like French onion soup, a slopping mélange of dark comb, amber honey, and frayed wax edges.

The uncapper's off-ramp is flanked on either side by waist-high stainless steel extracting tanks that measure about five feet across. Positioned between the ramp and an extractor, Keith pulls sticky uncapped frames off the belt two at a time, then loads them, short end first, into the brackets in the tank. When seventy-two frames are secured face to face within, the extractor lid is closed, the red control switch is flipped on, and the giant slinger begins its fifteen-minute spin cycle. As soon as one extractor is activated, a sticky-armed assistant moves to the other side of the uncapping ramp to load the second spinner.

In the rumbling extractor, honey is being slung out of the comb in glossy filaments like spun sugar in a cotton-candy machine. It drips thickly and slides down to the sloped bottom, oozing into a tub sunk into the floor between the two extractors. With a capacity of about a barrel of honey, the tub fills quickly. An icing of blond and brown stray wax caps conceals the honey beneath. Inches below this wax, a length of two-inch-wide white plastic piping sucks the honey up and delivers it to an elevated stainless steel holding tank in the front of the house.

The holding tank has a capacity of about eight barrels, or the proceeds from a large, prosperous beeyard. Metal partitions, which act as a filter for the wax that has collected in a thick, flexible crust on top, divide the volume into three sections. In the first compartment, the wax is deepest, forming a tough elastic skin that gives the honey the heavy, wavy bounce of a water bed. A wire basket with a handle, which looks as if it might be used to cook French fries at McDonald's, is periodically pulled across the top of the honey to collect and remove the wax. Through openings at the bottom of the tank partitions, honey seeps from the first segment of the tank into the second. Here, just a small amount of wax rises to the top to be skimmed off. The third part of the tub collects only the barest sheen of wax.

From this limpid amber pool, a spigot empties the honey into barrels rolled beneath the tank. The spout unleashes an inch-wide, muscular column of honey, a thing of surprising, luscious beauty, into the mouth of the waiting green canister. This is the river of reward for all the hard work of the season and the harvest day. Smiley usually does the barreling himself, savoring the loveliness and pride of the moment. Occasionally, just to show off, he'll tap the thick rope of honey as it exits the spigot. Like a snake, it is much drier than it appears. Due to its low moisture content (about 16.5 percent), the shiny column can be nudged and tapped, diverted briefly to one side, without spitting a lick of honey onto his finger.

Extracting the yield from a good-sized yard can take a full afternoon. The heat of the day pushes honey faster and more smoothly out of the comb and through the tanks and pipes, but the temperature inside the sun-baked metal house can climb to uncomfortable levels. A large round fan near the doors ineffectively blows the heat outside with a roar. With its hot sweet air and humming machinery, the honey house on harvest day feels like a busy, noisy bakery whose confectioners communicate by shouts and hand signals. I imagine that this delicious commotion is comparable to the inside of the hive during a busy nectar flow.

A late lunch is often a welcome break from the heat and excitement of the honey house. The meal might be a box of fried chicken from the gas station, sandwiches from Subway or home, or one of Smiley's Crock-Pot creations such as venison sausage and lima beans smothered in mayonnaise, wolfed down with Mountain Dew and Coke in the air-conditioned kitchen. George and Keith take a silent, satiated smoking break in the yard before heading back in to maneuver frames around the tight space in the busy honey house. When they are taken out of the extractor, the empty frames are windblown, light, and tacky with honey. The guys load them back into super boxes and cart them to the front door, opposite the holding tank. Then they tackle another stack of full boxes from the back of the house, dolly them over to the uncapping machine, and cycle them through. Frame by frame, hour by hour, the block of boxes that started the day in the beeyard gets processed, then rebuilt, empty, in the front of the house by evening. A few confused bees that never made it back to their hives scavenge off the sticky boards. Some of them fly across the room to inspect the barrels full of the day's freshly harvested tupelo.

The amount of the harvest is just about what Smiley had predicted that morning. Now, at the end of the day, he has to make more predictions. As he watches the luminous snake of honey flow into the last barrel, he's thinking about how much tupelo nectar is still out there, where it is, and how he's going to catch it. A recent heavy rain knocked the blooms off the trees, and the hot spell since then means that the flow will taper off quickly. He might have better luck switching to high gallberry bush and saw palmetto nectar, which are coming in strong and early. Every minute that he doesn't have all of his supers deployed to catch the strongest flow, he's losing money. Commercial beekeeping is a game of musical chairs with Mother Nature at the turntable.

Looking at the newly empty supers, he decides to split the difference, dividing the empties between his strongest tupelo yard and a yard where pearly white buds of high gallberry and butter yellow feathery fronds of saw palmetto have just begun to advertise their nectar. After dinner, he'll drive over and deposit more empty boxes onto the strongest tupelo flow. Then maybe he'll move the bees from the finished tupelo yards to the gallberry and palmetto, and leave the rest of the honey supers on those hives. He gives the signal to the boys to start loading the empty supers back onto the truck. It will go on like this all season, robbing hives in the morning, unleashing thick snakes of honey in the afternoon. In the evenings or very early mornings, when it's dusky and cool and the bees are inside, he moves the colonies to new nectar flows and places empty honey supers on top. Every day is a different gamble and a different challenge. "I like a challenge," he says, eyeing the stacks of supers. "If I didn't, there's no way I'd be doing this."

Smiley's challenges in calibrating and capturing his harvest are a recent historical development, although honeybees have produced their bounty in the wild for tens of millions of years. *Homo sapiens* arrived only about two million years ago, and, in a harsh nutritional landscape, began to scavenge and devour the precious sweet as often as he was lucky enough to find a rich vein of it tucked into a rock face or a tree cavity. If *Homo sapiens* and his descendants stumbled across a nest of bees in their daily wandering pursuit of food, they plundered it haphazardly and immediately, brushing away angry bees, enduring painful stings, prying out and gorging on the rare golden sweetness until all the honey was gone. For Smiley's ancient predecessors, honey was an opportunistic, orgiastic feast.

Millennia of subsistence food gathering passed, and honey remained an unexpected treat, a sweet blessing from nature and the gods. As humans became more settled and sophisticated, they still craved the treasure of the hive and sought to make it the reward of hunting rather than happenstance. They developed techniques and tools—a variety of ropes, harnesses, ladders, knives, and ways to

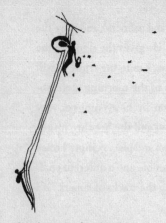

A rock painting found in the Arana Cave near Valencia in eastern Spain depicts honey hunting at around 6000 B.C.

transport and store the prize—to help find and rob the bees on a more regular basis. Near Valencia, Spain, a petroglyph from 6000 B.C. shows early honey hunters using some of these tools: A slim black figure on a tangle of lines explores a dark cavity at the top of a tall wall. One hand is deep in the hole, the other is grasping a vessel that looks like a gourd with a carrying strap. Winged creatures buzz around the hunter and the jar. Beneath all this activity, an additional figure is climbing the ropes, carrying another gourd in which to take away the combs.

A rock painting in central India 5,500 years later shows a similar, though much more elaborate, honey-hunting scene. At the center is a crescent-shaped slab of natural honeycomb, with black bee-dots traversing and actively filling the air around it. A ladder extending horizontally beneath supports a limber, delicately drawn human. This figure crouches next to the slab, grabbing it with one hand and with the other holding a receptacle beneath the comb. On a spindly vertical ladder to the right, a second figure dangles from the line. He wields an extended pronged tool, with which he seems to be spearing the comb and carving it into his comrade's waiting honey pot.

Honey harvesting continues this way today in many parts of Africa and Asia. A photograph that Eric Valli took of honey hunters in Nepal in the late 1980s brings ancient rock drawings to fantastic, colorful life. A sinewy ladder of green vines hangs down the side of a steep granite cliff face. Mani Lal, a sixty-four-year-old honey hunter, crouches on the bottom rung of the ladder, lean as the Indian petroglyph. He is stabbing a long bamboo pole into a slab of honeycomb attached to the rock face. A cloud of bees envelops him, clinging to his naked ankles and forearms, and to the burlap scarf draped over his head. As he carves the pregnant comb with his pole, it calves heavily into a nearby basket, which is supported by another set of ropes. Describing the scene in *National Geographic* magazine in November 1988, Valli wrote:

> He starts down the swaying ladder like a spider on a frail strand of web. The slightest error of judgment would mean death. Mani Lal stops beneath an overhang to face a nest nearly as large as he is. Its surface ripples with a thick black layer of bees. Two of the honey hunters, Krishna and Akam, have climbed a third of the way up the cliff to secure the ladder. Clinging to the rock, they pull the rope against the cliff to bring Mani Lal closer to the comb. Meanwhile a fire has been set at the base of the cliff to disorient the bees with smoke and encourage them to leave the nest. But the wind is blowing the smoke away. Gesturing toward the top of the cliff, Mani Lal issues a silent order. Within minutes a flaming bundles of leaves is lowered, and Mani Lal pushes it under the bees with a bamboo pole. Now panic runs over the living surface of the nest as the bees furiously depart in the smoke. Nothing distracts Mani Lal however. The golden comb has been unveiled. . . . By the time Mani Lal finishes filling the first basket, it's brimming with 15 litres of honeycomb. As it reaches the ground, villagers swoop upon it, dipping in bowls and breaking off small chunks of comb to chew.

After a lifetime in the forest, Mani Lal knows exactly where to find his feast each season. Historically, locating the nest was an arduous prelude to the sweet reward. Successful trackers watched flower blossoms dancing under the weight of feeding bees or observed bees sipping from streams or ponds, from which points they could be followed and traced back to their hideouts. Natural predators such as badgers and bears left tracks that hunters could conveniently follow to wild bees' nests. In parts of Africa, honey-loving birds, called honey guides, were enlisted to aid hunters in their quest. The birds, which are no bigger than a barn swallow, don't have the equipment or size to raid the nests themselves, so with a noisy, excited trail of chirps, they lead ratels (honey badgers) to a bee nest, relying on the intrepid mammal (and many humans) to do the dirty work for them and plunder the hive. Theodore Roosevelt wrote of the wily creature in 1910, while traveling in Africa:

> *Before starting for Africa John Burroughs had especially charged me to look personally into this extraordinary habit of the honey bird; a habit so extraordinary that he was inclined to disbelieve the reality of its existence. But it unquestionably does exist. Every experienced hunter and every native who lives in the wilderness has again and again been an eye-witness of it. Once while I was tracking game a honey bird made his appearance, chattering loudly and flying beside us; I let two of the porters follow it, and it led them to honey.... While camped on the Nzoi the honey birds were almost a nuisance; they were very common, and were continually accompanying us as we hunted, flying from tree to tree, and never ceasing their harsh chatter. Several times we followed*

*the birds, which in each case led to bee trees, and then perched
quietly by until the gun-bearers and porters got out the honey,
which we found excellent eating by the way.*

In exchange for the guide service, the honey bird expects a share
of the profits and is legendarily cruel if denied. Hans Schomburgk
described the wrath of the honey bird in 1910:

*The Zulus also know the honey bird, a small bird, which attracts the
attention of man by its twittering. Once it has been observed and is
being followed, it flies on and finally settles on a tree, within which
bees have made a nest. When the honey is removed, a part must be
left, otherwise, so the Zulus say, the next time the bird will guide
the man following it not to honey, but, in order to take revenge, to
a leopard or a black mamba, the most dangerous poisonous snake
of Africa.*

Lacking a honey guide, skilled listeners could determine if a
slight vibration or a distant hum meant a hive established nearby,
and a few experts claimed to be able to find nests aided by a trail of
tiny bee excrement. Most ingeniously, hunters used baits of honey
and water to attract bees so they could be tagged and followed back
to their homes. Columella described a method of tracking them in
his *De re Rustica* of the first century:

*First we must try to discover how far away they are, and for this
purpose liquid red-ochre must be prepared; then, after touching
the backs of the bees with this liquid as they are drinking, at the
spring, waiting in the same place you will be able more easily to*

recognize the bees when they return. If they are not slow in return-
ing, you will know that they dwell in the neighborhood.

While Columella preferred dots of vermilion, other ancient bee hunters used fluffy bits of feather, threads, or tender shoots of grass to mark and track the bees. If the quarry seemed to be living farther afield, Columella offered a more advanced solution, that of build-ing a bee trap:

> *A joint of a reed with the knots at either end is cut and a hole bored*
> *in the side of the rod thus formed through which you should drop*
> *a little honey. Then when a number of bees, attracted by the smell*
> *of the sweet liquid, have crept into it, the rod is taken away and the*
> *thumb placed on the hole and one bee only released at a time, which,*
> *when it has escaped, shows the line of flight to the observer.... Then,*
> *when he can no longer see the bee, he lets out another, and if it*
> *seeks the same quarter of the heavens he persists in following his*
> *former tracks.*

From red dots to reed tubes, hunters devised a number of meth-ods to contain and track their insect prey. In *Oak Openings*, written in 1848, James Fenimore Cooper depicts a bee hunter named Ben Boden, who was "extensively known throughout the Northwestern territories by the sobriquet of Ben Buzz," and describes his tech-nique of seducing and capturing bees with a bit of sweet honeycomb before following them back to their nest.

> *[Boden] found a bee to his mind, and watching the moment when*
> *the animal was sipping sweets from a head of white clover, he*

cautiously placed his blurred and green-looking tumbler over it, and made it his prisoner. The moment the bee found itself encircled with the glass, it took wing and attempted to rise. This carried it to the upper part of its prison, when Ben carefully introduced the unoccupied hand beneath the glass, and returned to the stump. Here he set the tumbler down on the platter in a way to bring the piece of honeycomb within its circle.

So much done successfully, and with very little trouble, Buzzing Ben examined his captive for a moment, to make sure that all was right. Then he took off his cap and placed it over tumbler, platter, honeycomb and bee. He now waited half a minute, when cautiously raising the cap again, it was seen that the bee, the moment a dark-ness like that of the hive came over it, had lighted on the comb, and commenced filling itself with the honey. When Ben took away the cap altogether, the head and half of the body of the bee was in one of the cells, its whole attention being bestowed on this unlooked-for hoard of treasure.

Using the overturned tumbler approach, Ben Buzz has several captives feasting on a piece of plattered comb in a matter of minutes. Removing the glass, he carries the plate like a waiter, watching care-fully as each liberated bee in succession takes flight and returns home. Following this bee relay to a tree some distance away, he pin-points the nest high in some decaying branches. Immediately the trunk is chopped down, the bees flee, and Buzz is rewarded with a small fortune of over 300 pounds of honey and wax.

Once they had located the nest, honey hunters seized their prize in radically different ways. In Africa and Australia, for example, it was often delicately swabbed out of the nest. A writer described the tribal methods used by aborigines of the Victoria River area in Australia: they

> *have invented an ingenious device by means of which they can secure honey from otherwise inaccessible fissures in rocks and hollows in stout-butted trees. A long stick is selected, to one end of which is tied a bundle of vegetable fibre or pounded bark. With the bundle forward, the stick is poked into the cleft leading to the hive, and, when the honeycomb is reached, it is turned around to absorb some of the honey. Then the stick is quickly removed and the absorbed honey squeezed from the fibres into a receptacle. The process is repeated, time after time, until the greater part of the honey has been obtained.*

The giant Q-tip technique and similar methods would have left the bees in peace, or at least alive. In Europe and North America, however, hunters like Ben Buzz considered bees to be a nuisance in endless supply, so they simply destroyed the nest in order to rob it, burning the henhouse to harvest the eggs. Those who weren't pillaging and destroying wild nests developed a more proprietary, caring relationship with their bees and honey. They realized, like the aborigines, that if the source of the treasure wasn't destroyed each time it was discovered, it could be husbanded and harvested again and again. By taking only the part of the nest that was dripping with honey, they left bees, brood, and food enough to produce further bounties.

Proprietary hunters guarded their honey spots as ferociously as anglers protect their best fishing holes. Honey men would claim "their" bees with piles of rocks in front of a nested cavity, or drawings of possession on the cliff face beneath. In tree nests, ropes tied around the trunks, bits of cloth nailed like flags, or slashes and burn marks

A magazine from 1881 portrays several ways to handle a swarm of bees.

were all used to brand the nest for its owner. In parts of eastern Europe, hunters peeled off a patch of bark on the tree and wrote their name, tattooing their deed and title on the tree.

As nests were revisited and husbanded, they became permanent resources for the beekeeper. In Europe until the twentieth century, the best bee trees were found deep in the woods, nestled close to the choicest nectar sources. Harvesters traveled there by foot; the routes to these spots were well-guarded secrets that came to be called "bee walks." Honey for sweetening and as a way to make the alcoholic beverage mead was so valuable (as was the wax used for making candles) that nests and bee walks themselves became a prized commodity, mentioned in many wills and deeds. On change of estate ownership, beekeepers who knew the walks were often transferred, too, their knowledge too prized to let go. The records of grand princes in fifteenth-century Russia note their parting with some property but retaining the rights to use the walks and visit the valuable bee trees.

Using a ladder or ropes, or even pegs or notches carved into the tree, the harvester could visit his wild warehouse several times during the season. Nests had a small front entrance for the bees and a man-made back door through which the harvester could smoke and gain access to his crop. On each visit, he would pry open the door, smoke the colony, and then carefully remove a few of the outermost combs, which in a typical nest formation would have been full of stored honey and mostly free of brood. Resealing the door, he would wait a few weeks or months for the bees to replace and refill the combs before visiting and plundering again. Varieties of wild nest management were practiced all over the world, in all manner of shape, size, and innovation. In central Africa, hives were kept in hollow wooden cylinders suspended in trees close to the original nest. At

Northern European tree beekeepers at work in the 1700s.

harvest time, the nest could be lowered, the combs extracted, and the hive closed and pulleyed back to its original spot.

Most likely first in Egypt, ancient hunters lost patience with the distance, effort, and time involved in finding or maintaining nests in the wild. Temple priests could procure their sweetener more conveniently and reliably by relocating nests into the backyard. Discovering that bees happily set up housekeeping in any container that would accommodate their numbers and keep them warm and dry, Egyptians fashioned clay vessels to capture, transport, and house bees. Diamond-shaped containers, resembling two giant pie plates joined rim to rim, were smeared inside with honey and fragrant herbs and placed in the wild in hopes of attracting a swarm. Sometimes larger pottery containers the size and shape of a watermelon were used. If a colony settled within, it was brought back to the temple to join the apiary. Thus, in some parts of the world, honey hunting evolved very early on into domesticated beekeeping.

Domesticated is really the wrong word, though it is so often used. Bees are gentle and accommodating but never yoked, trained, or housebroken. No matter where they are kept (another inappropriate word), they are wild and untamed in their instincts and actions. In some ways bees "keep" us, humoring us with honey and their divine presence if we stay on our best host behavior. The Egyptian priests seem to have sensed this and harvested only the laden honeycomb, leaving the bees and brood to thrive. By the second millennium B.C. in Egypt, beekeeping had become a sophisticated art. A wall painting from this period, in the tomb of Rekhmire in Luxor, shows two beekeepers crouched in front of a stack of oblong hives. One attendant is smoking the tower, while another is removing combs. A bee flies amiably between the two.

Wall paintings from the Eighteenth Dynasty Tomb of Rekhmire
depict the gathering of sacred honey.

The science and craft of hive beekeeping spread from Egypt to
ancient Greece and Rome, with beekeepers fashioning homes for
the bees out of whatever materials were locally available. In Egypt
this was clay; in Europe it was wood and reeds. Hives in the Mediterranean were most often large buckets woven of osiers or cylinders assembled of cork, in which, under a protective lid of clay or
straw, the comb hung down from sticks suspended across the open
upturned mouth. Columella offered the following tips on materials
and construction in the first century:

> *Beehives must be constructed in accordance with local condi-*
> *tions. If the place is rich in cork-trees, we shall certainly make*
> *the most serviceable hives from their bark...or if it grows*

The proud owner of some tree-stump hives.
And some without proud owners present.

> *plenty of fennel stalks, ... receptacles can be quite as conveniently*
> *made by weaving them together. If neither of these materials is at*
> *hand, the hives can be made by plaiting withies [flexible twigs]*
> *together; or, ... made with wood of a tree either hollow or cut up*
> *into boards.*

For the next seventeen centuries, hive beekeeping remained fairly constant in Egypt and Greece in practice and vessel. In northern Europe and the United States, beekeeping followed the tradition of using whatever materials were most readily available and comfortable for the bees. Nests were most often found in trees, so beekeepers simply chopped down the occupied part and relocated it to the beeyard. Organized apiaries started as haphazard congregations of transplanted sticky stumps.

As these natural hollows degraded with time, the logs and limbs were replaced with man-made vessels of straw, wood, cork, or other materials cheaply and conveniently at hand. Bees thrived in all kinds and sizes of vessel, and apiaries were assemblies of clay pots, buckets, gourds, logs, even cookie tins. German and Italian beekeepers carved elaborate wooden statues of saints and folk icons complete with nesting cavities and entrances for bees. All were adapted as homes by the accommodating guests.

The most standard, popular man-made hives were called skeps, which began as overturned baskets and evolved into plump rounded domes of coiled straw or reeds resembling cake

A sixteenth-century woodcut shows two skeps, a popular type of man-made bee house, and the equipment worn to visit them.

domes or miniature igloos, complete with an entrance at the base. These were easy and cheap to make, and most European villages had an expert skeppist or two in case no one in the family was inclined or skilled enough to make them. In *Beekeeping New and Old*, published in 1930, William Herrod-Hempsall describes the skep tradition in England, where he grew up:

> *The farm labourer or hind of years gone by made his own straw hives during the long winter evenings. Our maternal grandfather used to do this over one hundred years ago. Rye straw being the most favoured material, on account of its length and toughness. Wheat straw was also employed, the farmer allowing his men to take the straw without payment.... The butt end of a bullocks horn cut off about one and a half inches wide and well polished, formed a gauge to draw the straw band through, so that all the coils in the skep were of equal thickness, these rings were handed down from father to son as heirlooms.*

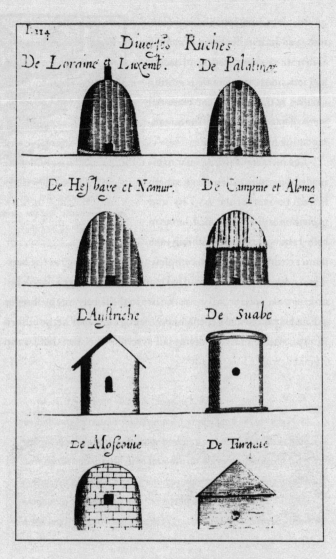

A 1646 drawing of the various materials and constructions
of beehives throughout Europe.

Beehives were housed on garden floors, on tables, in stone wall niches or apiary platforms, and during the nectar season bees foraged at their leisure and returned to the skep to fill the comb they had attached to the rounded ceiling. Deluxe skeps were made with handles, so the contents could be easily managed, moved, and taken to market (as long as their bottoms were tied off with cloth).

Its lightweight portability meant that the skep was versatile and adaptable to many different harvesting techniques. Charles Butler, an Englishman, wrote *The Feminine Monarchie* in 1609, and described therein the state of the art. The first was simply "driving" the bees from their laden skep into an empty one. This was accomplished by laying the occupied colony on its side, with a wooden gangplank or ramp extending from its mouth into another that had been temptingly smeared with honey and herbs. By shaking, tapping, smoking, making loud noises, and generally irritating the bees, the beekeeper persuaded his stock to flee their home and seek shelter in the calmer empty basket. Their abandoned honeycombs were then left for the

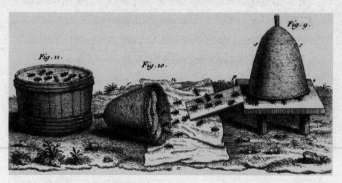

A plate from René-Antoine de Réaumur's 1742 *Mémoires pour servir a l'histoire des insectes* (Natural History of Insects) shows drowning and driving, two popular methods of getting the bees away from their honey for harvest.

delectation of the beekeeper. Butler suggested this method for use in the late fall, when there was very little brood in the colony, as bees could not otherwise be induced to leave their young, but he and many of his countrymen considered driving the bees to be cumbersome, time-consuming, and painful. It was not his first choice.

The second method set forth in *The Feminine Monarchie*, descended from the Egyptians and Greeks to generations of nest beekeepers—which got honey onto the tongue and hearth much faster—is referred to as "castration." In this practice, the outer honeycombs were cut from the hive, and the inner brood comb and bees were left for future production. On a visit to Greece in 1675, another Englishman, George Wheler, described the castration, which hadn't changed much since Columella's day:

> The hives they keep their bees in are made of willows or osiers, fashioned like our common dustbaskets, wide at the top and narrow at the bottom, and plastered with clay or loam within and without... The tops are covered with broad flat sticks, which are also plastered over with clay.... Along each of these sticks, the bees fasten their combs; so that a comb may be taken out whole, ...with the greatest ease imaginable. In August they take out their honey, which they do in the day time also, while they are abroad; the bees being thereby, say they, disturbed least; at which time they take out the combs laden with honey... beginning at each outside, and so taking away, until they have left only such quantity of combs, in the middle, as they judge will be sufficient to maintain the bees in winter.

Although this method had been practiced successfully for centuries, Butler does not seem to believe it really works. He concludes that reducing the nest in this way "surely must needs do more harm than good." Like Ben Buzz, Butler's preference was for death, which he promoted eagerly. "The most usual and generally most useful manner of taking the combs is by killing the bees," he declared, reflecting the refined ignorance and lazy beekeeping of the day. *The Feminine Monarchie* counsels the beekeeper to choose the oldest, weakest colonies in the yard to be killed and harvested, leaving the rest to survive for the next season. Then, "Having made your choice of your stalls [skeps] to be taken, from two to three hours before sunset, dig a hole in the ground about eight or nine inches deep and almost as wide as the hive skirts." In cheerful detail, Butler advises lighting a smoking sulfur fire in the hole and clamping the skep down over the pit, fatally trapping the bees in the dome of noxious fumes. "So shall you have the bees dead and down in less than a quarter of an hour." After a few minutes, the dead bodies could be shaken out of the skep and the honey harvested without interference. For those who preferred asphyxiation to poisoning, he recommended plunging the skep into a tub of water to drown the bees and rid the vessel of its irritating occupants. This killing and plundering was popular, widespread honey practice for centuries. Though the honey was handled with great care, many European keepers considered bees highly dispensable, abundant pests best eliminated once their honey had been delivered, a far cry from the reverence in which the ancients had held the bee.

Although killing the bees was the most dastardly, all three popular methods involved damage to the hive because harvesters had to cut combs—destructively and irreversibly—from the nest. Castration was akin to lopping off laden branches from a healthy apple tree, causing injury and requiring time to regenerate the branch and fruit. The killing method was like picking all the apples off a tree before savagely chopping it to the ground. By the 1700s, however, more and more European keepers were thinking about how to get the most fruit with the least amount of damage to the tree, and the killing trend slowed. Periodic shortages of bees emphasized the radical notion that perhaps it would be good to emulate the Egyptians and the Greeks and nurture the bees rather than murder them.

Thomas Wildman was one of the enlightened beekeepers. In 1748 he wrote: "Were we to kill the hen for her egg, the cow for her milk, or the sheep for the fleece it bears, every one would instantly see how much we should act contrary to our own interest: and yet this is practiced every year, in our inhuman and impolitic slaughter of the bees. Would it not argue more wisdom in us, to be contented with taking away only a portion of their wax and honey, as is the practice of many countries."

Beekeepers gradually discovered the intimate workings of the hive and realized that bees could be a renewable, breedable resource. As colonies of bees became longer-lived investments, so did their homes. Many European bees moved out of straw skeps and disheveled logs into more elaborate, permanent wooden boxes. Bees and their belongings were increasingly housed as honored guests in the backyard, garden, or orchard, and given permanent wooden dwellings with all the modern conveniences, some even with delicate gingerbread trim and mansard roofs.

Honey enthusiasts also learned that bees, in their constant zeal for surplus, would stockpile honey, as much as they were given room for. Annexes of wood, straw, clay, or even glass were placed adjacent to the brood colonies, and the bees instinctively stored excess honey in them. Those placed on top of the hive came to be

A nineteenth-century drawing of the octagonal Stewarton hive.

called supers, the ancestors of Smiley's hundreds, because of their position above the brood colony. When the extension was full, the harvester removed it, smoked and brushed it, or dunked the super section in water until the bees were dazed enough to relinquish their surplus food supply. In "Hints for Promoting a Bee Society," Dr. John Lettsom wrote, in 1796, "Plunge the whole box and its contents in a tub of water, placed in readiness for the purpose. Neither the wax nor honey will be harmed if done with a gentle hand, and not immersed too long; and the bees will soon recover their drowning, if taken out and laid on a dry cloth in the sun."

Although many farmers and beekeepers in Europe remained loyal to simple straw skeps, others had contraptions for housing bees and capturing honey that were as varied and elaborate as birdcages and doghouses. Early beekeeping catalogues and broadsides depict a range of sizes, shapes, and complexity, each touted as the best method ever for accommodating bees and easily relieving them of their surplus.

As in any cultish pastime, a variety of gear was involved, with varying opinions on how to use it to obtain optimum results. By the 1800s, glass bell jar caps atop the hive had become popular in England. R. Hoy wrote in *Proper Directions for How to Manage Bees in Hoy's Octagon Box Bee Hives* in 1788:

> *When your glasses are filled with honey, and sealed up, take them off: if they should be fastened, run a thin knife under them, and take them away some distance from the hive. Then sweep the bees off with a feather as they come up the glass, and they will fly home to their habitation. Then, if neither the frost nor damp gets at them, tie your glasses over with paper, to prevent the bees from robbing them of the honey; they will keep good in glasses two years.*

Even with increased, managed surpluses, the limitations of the hive supers meant that honey was still a small local harvest. One super, or one bell jar at a time, the comb was cut from the surplus layer, bees were removed, and then the honey was consumed or sold in its original beeswax packaging. To extract "liquid" honey, the whole comb could be crushed and wrung by hand or dumped into a screening cloth, then squeezed, twisted, and hung until the sweet liquid guts oozed out. Sometimes the comb was heated to melt the wax, which was then lifted, cooled, off the top of the liquid honey. No matter how it got to the table, obtained from a local beekeeper or the backyard hives, honey was a messy, waxy, impure production. Whether three hives or thirty were gathered in the yard, in elaborate painted dollhouses or plain old skeps, the harvest required a maximum of patience and effort for a minimal, delicate crop. By the beginning of the nineteenth century, beekeepers were inspired by impatience, greed, and curiosity to find a way to steal more honey more easily.

The leap from cottage to commercial industry occurred in the middle of that century, when Langstroth's bee-space breakthrough meant that frames of honeycomb could be easily and harmlessly taken from the hive. In America and Europe, a cluster of inventions followed that soon solved the remaining sticky problem of getting the honey out of the comb. In 1865 an Italian beekeeper named Major Franz Edler von Hruschka helped reduce the mess by inventing the *smielatore*, or honey extractor. The third edition of *Langstroth on the Hive and the Honey Bee*, published in 1889, tells the story, which many consider apocryphal but which is too charming to omit: "It happened in this wise: He had given to his son a small piece of comb honey on a plate. The boy put the plate in his basket, and swung the basket around him, like a sling. Hruschka noticed that some honey had been drained out by the motion, and concluded that combs could be emptied by centrifugal force." Some say it was the boy's buxom maid who swung the basket, but either way, it got Hruschka's attention. He devised a crank-driven contraption that secured the frames and slung them around as if in a giant basket or a modern-day washing machine. First, he cut the caps off the comb with a knife, so that the force of the slinger effortlessly and neatly flung the liquid from its wax casing. Pure honey could be poured from the extractor into buckets and jars for storage or sale, and the empty combs, fully intact, could then be replaced in the hive for immediate refill, increasing yields exponentially. The editors of *Langstroth on the Hive and the Honey Bee* wrote: "This invention was hailed, in the whole beekeeping world, as equal to, and the complement of, the invention of movable frames." All over the world, a flurry of honey "slinger" inventions followed Hruschka's. In 1869, H. O. Peabody of Boston applied for a patent on his hand-cranked extractor, which was the first to be manufactured for sale and distributed in the United States. A better extraction method, if there is one, has yet to be discovered. Hobbyist beekeepers today still extract their honey with old-fashioned centrifugal force and elbow grease, cranking the handle of a stainless-steel drum appliance that spins honey from two, four, six, or eight frames of comb. The commercial-strength extractors that Smiley uses in Wewahitchka are electric and bigger, but otherwise not much different from Peabody's invention.

Having observed the amount of time, energy, and nectar resources that the bees put into building their comb from scratch with wax they had laboriously secreted from abdominal glands, John Mehring thought to aid them (and increase honey output) by providing wax foundations for the comb, as if donating prefabricated walls or furnished rooms to the bees. Mehring carved precise hexagonal patterns into two sheets of wood and pressed thin layers of wax between them. Placing the resultant sheets in the hive, he found that the bees adopted it readily and productively, happy to have the construction help. These first wax sheets were found to be too delicate for the violence of the modern honey slingers, so J. F. Hetherington, an American, successfully reinforced them with wires, which allowed bees to build and fill the comb, and farmers to extract it, at a much faster rate. Thanks to the *smielatore* and J. F. Hetherington, most of the comb that Smiley uses has been filled, emptied, and reused through several harvests.

Smiley's smoker is also the result of nineteenth-century innovations. Versions of smoke and smokers had been used since the earliest apiaries. The first, adopted after beekeepers observed the evacuative effects of fire on a bee nest, were probably torches waved in front of the hive. The Egyptians upgraded to a bowl containing smoking embers. Though the smoker's fuel—wood chips, pine needles, straw, even tobacco—varied according to region and convenience, the simple smoking bowl did not change very much over the centuries.

1889 advertisements for some of the new equipment and Quinby's book on beekeeping.

The introduction and sophistication of movable frame hives in the late 1850s demanded a new, more delicate type of smoker that could provide a relatively small, gentle gust to calm the bees; it was no longer desirable to produce a giant fiery plume of acrid smoke to evict them. A variety of contraptions and experimental gusts of smoke were employed from the 1850s to the 1870s. An American, Moses Quinby, gets official credit for first attaching a bellows to a handheld tin can of smoke in 1870, a prototype of the one most beekeepers use today.

Movable frames, extractors, prefab foundations, and precise, gentle smokers allowed beekeepers to minimize labor, stings, and bee death while harvesting rivers of honey. By the late nineteenth century the savage opportunism and destruction of man's earlier relationship with honeybees had evolved into a civilized partnership. Since this collaboration began, bees have generously produced as much excess honey as seasons, flowers, frames, and good management have allowed.

The partnership soon became refined and confident enough that beekeepers even began experimenting with the type of bees they employed. A Swiss beekeeper, Captain Baldenstein, stationed in Italy during the Napoleonic Wars, had observed the complacency, productivity, and overall good manners of the local bees. After the war, he sent emissaries to Italy to purchase some of the natives and successfully raised colonies of gentle, generous Italians, writing of their supremacy in European bee magazines. Word of the new bees made its way to the United States, where Lorenzo Langstroth, despite his head troubles and retiring nature, was soon involved in ambitious importation schemes.

The first successful American immigration of these Italians appears to have occurred in 1860, when S. B. Parsons, acting as an agent for the U.S. government, bought twenty colonies of bees in Italy and returned with them by steamer to New York City. The royal passengers were unceremoniously stowed in cigar boxes with only a wedge of honeycomb and a small entourage of helpers, and only two survived the difficult crossing. These two proliferated and became the probable ancestresses of millions of Italian bees living and working in America today. New bee strains are always being tested and applauded, but most modern honey farmers and hobbyists use the Italians—their reputation for gentleness, and for storing excessive amounts of honey, has proven a lucrative discovery for generations of American beekeepers, including Don Smiley. Although at one point he experimented with Carniolans, a Slovene breed, his hives are now 100 percent staffed by Italians. "They store more honey," he says of his European business partners. "And they're just sweeter to work with."

Some Italians are still milling dazedly around the honey house at the end of the harvest day. A few are investigating the sticky interior of the extractor, the gentle, shining result of hundreds of years of harvest frustration and human innovation. The strays might leave the honey house and hitch a ride on the load of sticky empty supers that Smiley takes back to the tupelo yard that night after dinner. From there, he will transfer some of the colonies to a new gallberry yard before it gets too dark and he's finally tired enough to call it quits for the day. The next day, he'll do it all over again, and so will the bees. Though their pantry has been smoked, acidly fumigated, raided, and carted off, and their home has been relocated overnight, the sweet Italians will venture cheerily forth, reorient themselves, and begin to collect a new nectar in tireless service to nature, the hive, and the harvest.

RESPECT

(Or, never underestimate your bees.)

By my second season, I had doubled my operation to two hives. The weather had cooperated, and by October I had over thirty heavily full frames ready for extraction. My super frames were five inches deep, and each now weighed about five pounds, so my sweet haul was looking very promising. I had borrowed Ace's hand-cranked four-frame extractor for the job and had even ordered glass jars in which I would proudly display and distribute my crop. On the day of the harvest I realized that my greed and gloating enthusiasm had eclipsed my thorough planning, because I was lacking a bee escape or butyric acid, the means of getting the bees off their honey in order to extract it.

I was forced to remove the bees by a sloppy, laborious, and comic combination of smoking and brushing each frame, which was kind of like licking a melting ice cream cone on a sweltering day. I would smoke and brush one side of a frame, then turn it over to repeat the process, by which time the first side was again dripping with bees. I smoked them so much that the frames began to smell like wood smoke and I wondered if the honey inside would taste like bacon. Finally I would get it down to just a few bees and then run away from the hive, each time escorted down the hill by a small, persistent contingent trying to reclaim its precious surplus. I rested the frames upright like skis against the walls outside my honey house (the glassed-in back entryway of my house) because there were still too many bees to bring them inside. By the time I got the frames mostly debee'd and down the hill, hours later, I had a swirling encampment of bees outside my back door and had to enter the house through another entrance to avoid too many curious, unwanted insect guests. Dusk was settling in by the time I had finished the work that butyric acid would have accomplished in five minutes, and there was still an embarrassing cloud of bees outside, so I decided to extract in the morning. I figured that the loiterers would get cold, lose interest, and go home during the night, leaving me to rob them in peace the next day.

The following morning, the frames looked different. The night before, they had been heavy and full and sealed in a smooth coat of wax. The comb surfaces were now ragged and frayed, as if mice had been at work nibbling the surface. Looking closely, I saw that a quarter of the cells were completely empty of honey. Into the previous dusk, or even early that morning, the bees had industriously pried off the wax caps and emptied as many hexagons as they could, dragging the loot back up the hill to the hive. In a very sporting and appropriate double cross, the bees had robbed me. I had gravely underestimated the instinctive cunning of the bees and their understandable attachment to their hard-earned honey. My beekeeping humility, and my admiration for the bees, has never been greater. I extracted my diminished prize, and savored it more.

FOUR

Pollination

*Without this total contribution by bees, we would certainly live
in a very different, less productive and less interesting world.*
From *Langstroth on the Hive and the Honey Bee*

*The brilliant shapes and colors of flowers (what more precise
record could there be of the esthetic preferences of bees?).*
Frederick Turner

*To make a prairie it takes a clover and one bee,
One clover, and a bee, and revery.*
Emily Dickinson

T wo hundred million years ago, bees did not exist. Their ancestors, a large, unruly family of wasps by the name of *Hyme-
noptera*, lived in a humanless, flowerless landscape in Asia. Members
of this carnivorous clan were fierce, predatory hunters, stinging,
killing, and scavenging to provide food for their young. When
flowering plants appeared fifty million years later, some of these

wasps abandoned the rapacious life to take peaceful advantage of the newly abundant food source. To gather plant pollen, the new nourishment, they developed hairs and collection baskets on their bodies. They cultivated long tongues to suck nectar and honey sacs in which to carry it home. The new creatures also developed gentler dispositions; unlike their warmongering wasp cousins, the bee-ish wasps were pacifists that stung only in self-defense and fed their young a strictly vegetarian diet. (Because of the family resemblance, bees are still often confused with wasps and suffer unfairly from their bad, thuggish reputation.) By the Cenozoic era, fifty million years ago, gentle bees as we know them had evolved, born out of their zeal for plant nectar and pollen.

Flowering plants are a prerequisite for the bee, and vice versa. Searching for food for their young, bees track pollen from one flower to the next, fertilizing the plants en route and enabling them to reproduce and bear fruit. Very recently, in the last hundred years or so, beekeepers joined this ancient coevolutionary partnership by renting their bees out for crop pollination. Smiley is just getting started in the business, renting his bees to cucumber, squash, and watermelon growers. In this happy arrangement, the beekeeper boosts his income, bees enjoy feasts of nectar and pollen, plants get fertilized, crops and farmers flourish, and summer picnics are blessed with watermelon.

Bees and watermelons arrived in North America together. The luscious fruit was first recorded in Massachusetts in 1629, nine years after the Pilgrims arrived with seeds and bees at Plymouth Rock. Settlers, seeds, and insects moved slowly south, and by 1664 there were watermelons in Florida. Centuries later, they are a $60 million crop in the state, the largest in the nation, and entirely dependent on bees for pollination. We have bees to thank for the bright pink joy of watermelons (and the leafy torture of the Brussels sprouts we were forced to eat as kids).

By the second week in May, the tupelo flow is over, and Smiley has moved all of his hives to gallberry patches, where the bees are scrambling about on clusters of pearly white buds. About sixty hives are earmarked for pollination, sitting in a defunct tupelo yard, while Smiley awaits a call from the watermelon man telling him where and when he needs the bees. By using the pollinating services of bees, cultivators can boost their yields by as much as 60 percent, yet only a third of Florida farmers report hiring them. "Some farmers around here don't really understand it yet," Smiley says. "They don't understand that everything benefits from insect pollination, and the more bees they have, the better."

On the designated delivery day, Smiley gets up at around 4:00 A.M. and is out of the house and in his truck half an hour later. Sleepily sipping a cup of coffee, he pulls into a darkened yard and turns off the ignition, leaving the headlights beaming onto rows of single white hive boxes. These nests are without honey supers, for there is no more nectar to be gathered from the nearby tupelo trees. The bees inside have been energetically building brood all spring and now are intent on feeding the larvae. By reassigning these boxes to a blossoming crop of watermelons an hour north in Chipley, Smiley is taking them to a pollen feast. During their long working vacation, his bees will have all the food they can gather, and they'll cross-pollinate the watermelons and earn him a thirty-five-dollar fee per hive. As long as the tupelo is over, it's a nice way to supplement his honey business. "Last year I wouldn't go to pollination because the tupelo flow was so good I couldn't leave it," he explains. But otherwise, "It's easy money. You drive up there, set 'em down, and that's it. If you make any honey, that's a plus."

He arrives at the watermelon field at around 7:00 A.M., long after fingers of orange sunlight have gripped the horizon, revealing a shadowy carpet of broad leaves and hairy vines. On one side of the forty-acre plot are the hives he brought up a week earlier, when the farmer first called him. "He tells me where to put them and then stands waaaaaaay back," laughs Smiley. Along the opposite side of the field, he arranges the new arrivals in a neat white row, with entrances facing the crop. Bestowing a honey super on each, he's finished with his part of the pollination business.

Returning to his truck, he swigs coffee and heads back down south to Wewa. He'll be home by nine or so and plans to spend the day bottling the new harvest of tupelo honey. Because of all the rain, he brought in just fifty barrels, half of what he harvested the year before. But that was a bumper year. "Every ten years you'll have a bumper crop," he says. "We had two years of bumper crop in a row in 2000 and 2001, so I guess we're done for the decade." He and Paula sell it, and every year sell out at the tupelo festival held in Wewahitchka in late May. The sixty-year-strong event draws thousands of visitors to Lake Alice park in town to sample live music, crafts, boiled peanuts, fried dough, and tupelo honey. There is inevitably a throng of people gathered around the table he sets up next to his truck, eager to taste and purchase his trademark. He figures he'll need to fill about a thousand twelve-ounce bears to satisfy his festival customers.

As Smiley arrives back in Wewa an hour or two after sunrise, the greenish yellow watermelon blossoms in Chipley flex open. Flowers produce and display their pollen first thing in the morning, like bakers setting out fresh, aromatic pastries. When the sun has warmed the hives enough for the bees to venture out, they catch the scent and dart eagerly from the entrance to the nearest yellow purveyor, attracted by its fresh aroma. A pair of antennae extending from the bee's head acts as her nose, an organ one hundred times more sensitive than the human version, especially when it comes to floral sweets.

Near the base of her antennae, the bee has a collection of eyes. Three simple ones, or ocelli, on the top of her head navigate the dimness of the hive. Outside, she employs two bulging compound eyes, perched on her head like a pair of dark wraparound sunglasses, which are composed of thousands of rounded hexagonal facets that sense light, color, contrast, and even movement as a prismatic mosaic of color and shape. These lidless eyes are always open and active, alert to the attractions of plants and nectar. As she approaches a fragrant blossom, the bee focuses on its nectar guides, a series of spots, dots, and stripes in the flower's interior that act, along with the petal and pistil arrangement, as arrows and billboards advertising its hidden treasures. The nectar guides do not reflect ultraviolet light the way the surfaces on the rest of the plant do, so the bees, which see ultraviolet very well, discern the guides as black contrasts against a pale floral background, the inverse of an airport runway at night.

Landing inside a blossom, some bees go directly to the well-signed nectaries to gather sugary sap for the colony. During the summer season, when the brood is numerous and clamoring, a third of the foraging force is obsessed with pollen, the male reproductive material of the plant that is fed to the bee larvae. Unlike most other insects, bees feed their young with these granules of rich plant protein. Its nitrogen, phosphorus, amino acids, and an impressively complete array of vitamins make plant pollen a perfect infant formula for the growing vegetarian bee. The nectar is pure carbohydrate, energy for adults, brood, and humans alike.

A pollen collector swoops determinedly into the center of the watermelon blossom, clambering around the swampy center with six searching feet. Each foot can taste what it touches, so as the bee dances busily around the interior, she uses all of her limbs to grope and sample her target. Advancing on the three towering stamens, the male sex organs, whose swollen tips are adorned with newly secreted pollen grains, she selects one and goes aggressively to work with her mouth and mandibles, massaging and loosening the gluey yellow particles. Quickly she transfers her prize from her jaw and forefeet to her middle feet. From there she relays it to her back legs, where two collection baskets, tiny saddlebags called "corbiculae," are ready to receive and store the load.

As she manipulates the stamen, loosened pollen grains rain down on her like ripe fruit from a shaken tree. Sticky pollen from neighboring stamens showers down as well, clinging like mist to the thousands of hairs covering her body. She grooms herself constantly, collecting and organizing this cloud of protein. Each of her limbs is outfitted with spiny comblike hairs that gather and guide the pollen into the baskets as she rakes her body in downward sweeps. This bodily collection method is aided by static electricity, which bees create in flight, up to an equivalent of 450 volts, or about four times household current. Inside the well-insulated blossom, pollen grains actually jump onto the fur of the highly charged bee, static cling at its best. When she has completed her gathering, in about six or seven seconds, her saddlebags are partially full, and she is coated with an electrified confetti of sticky male plant matter.

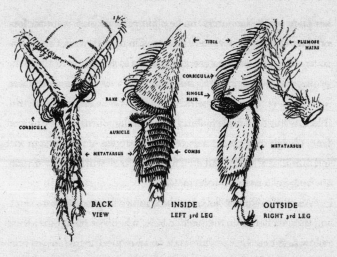

The phenomenal pollen-collecting architecture of a worker bee's hind legs.

As soon as she leaves one watermelon blossom, she searches greedily for another, guided by its scent and markings. Bees focus exclusively on one type of flower each time they embark on a foraging trip, so if a collection journey begins on dandelion or tulip or apple, it will end likewise. En route to her next watermelon, she continues to organize her bounty, brushing her body by using the grooming aids built into her limbs. Scraping her airborne hindmost legs together, she strokes and pushes the collected pollen into neat little pellets. She adds a dab of nectar to form a malleable mud, then bends and pumps her legs to compact the pollen into the collection baskets. The smooth indented face and spiked sides of the corbiculae keep the precious load in place as the bee shops from blossom to blossom, adding more watermelon baby food to her cart at every stop.

No matter how expert her in-flight grooming skills are, thousands of uncombed and uncollected grains of pollen still cling to her body when she enters the next blossom. Because watermelons produce one female for every six or seven male flowers, the pollen-covered bee will eventually enter a female blossom to investigate the protein resources. Finding none, she will exit only seconds later. The bee is perhaps disappointed, but the female flower is satisfied. The few hundred male pollen grains that have brushed or dropped from the bee onto the stickily receptive stigmas of the blossom will fertilize it. If all goes well, in a few weeks she will proudly display the bulge of a nascent watermelon.

A pollen grain that falls onto the stigma will grow a tube descending from its height to the swollen base, where the plant ovaries are encased. When this conduit lands on an ovule, the two sperm cells of the pollen grain slide down onto it. One fertilizes the egg to become a seed, while the other develops into tissue, food for the embryo. The pink, porous flesh of a watermelon is the food supply for each of the hundreds of fertilized seeds within.

Watermelon flowers have 1,000 ovules arranged on three lobes. Every one of these eggs must be fertilized to create flesh for a shapely, desirable fruit. Each bee that visits, with her furry aura of pollen, introduces, hopefully, hundreds of male cells to the expectant female flower. For adequate fertilization, each watermelon blossom requires at least eight pollen-coated visitors. Adding to the drama, her callers must all arrive in the few short hours after her petals open, because they will close permanently at the end of one day.

After visiting between ten and a hundred watermelon flowers, the bee has left a messy trail of fertilizing pollen at each stop. She has also assembled a dense little pellet of nutrition on her rear legs and returns home to unload her cargo. Her journey in the watermelon field has yielded a loaf of pollen the size of a third of a grain of instant rice, weighing as much as one third of her body weight. She looks as if she is wearing colorful inflated water wings or miniature orange pontoons on her rear legs, but they don't seem to help her buoyancy or balance. Wobbly and slow on return to the hive, she is hampered by her burden. Fifty thousand of these shipments are brought into the hive every day. Every year, a strong colony rears about 200,000 bees, nurtured by millions of tiny loaves, or up to seventy-five pounds of pollen.

Arriving back at the dark bustle of the hive, the laden bee looks around for a cell where she can unburden herself. She chooses one that is in close proximity to the brood cells, a convenience for the other bees working in the nursery. Finding an appropriate one, she grabs it with her front legs and perches her abdomen on the opposite rim so that her four other limbs dangle into the waxen well. Rubbing these together, she combs and scrapes the contents of her corbiculae into the storage bin below. Delivery complete, she dismounts and wanders back toward the hive entrance, readying herself for the next sortie. Eventually another worker bee will arrive to scrape and lower her plunder, adding to the babies' food supply. A typical brood frame pulled from a hive has a half-moon-shaped nursery of bee larvae, ringed with a wide band of cells containing brightly colored stores of pollen.

Soon, a nurse bee will scoop a portion of this collected pollen from the ring onto an egg or milky white, fat larva in an adjacent brood cell. Each nascent bee receives an estimated 10,000 meals during the eight or nine days of its egg and larval stages. With a snack required every few minutes, the nursery is in a constant feeding frenzy. On this nonstop nutritional formula of pollen and honey from nearby cells, the ravenous infant bee is on one continuous growth spurt, multiplying its bulk every day. After about three weeks, pupae burst from their cribs and begin the adult work of the hive. Without pollen, this next generation of bees would not grow to hatch, and the entire community would soon expire.

With instinctively urgent purpose, a bee makes up to fifty pollen-collecting trips a day, determinedly gathering food for the young and unwittingly fertilizing the crop. Each trip lasts from five minutes to a couple of hours, depending on the proximity and abundance of the pollen. Tirelessly, she'll gather and spread food, dabbling and darting through the sex life of plants until sunlight and temperature start to wane, forcing her back to the hive. Flowers that did not receive enough seminal visitors in their one day of blossom will abort and eventually drop from the vine. The lucky floral ladies that welcomed an adequate number of callers will produce a luscious round watermelon in about ninety days. Each succulent fruit is an unlikely triumph of timing, sex, and bees.

Aristotle was the first to write about this triumph, though he didn't know what it was. In *Historia Animalium*, he observed that "The bees pick up the wax (which was actually pollen) by scrabbling at the blossoms busily with their front feet; these they wipe off on the middle feet, and the middle ones onto the bent parts of the hind ones; having done this they fly away carrying the load and are clearly weighed down." He also described the floral allegiance of bees in their quest for food, the core of cross-pollination, without grasping its implications. "On each flight the bee does not go on to flowers different in form; it goes for example from violet to violet, and does not touch any other before it has flown back to the hive."

His student Theophrastus compiled a catalog of the known natural world in the fourth century B.C. entitled *Enquiry into Plants*. He observed that in many cases there were separate male and female versions of a plant (he based his assignments of sex mostly on their appearance), and that the two needed help in meeting and mating.

> *With dates, it is helpful to bring the male to the female; for it is the male which causes the fruit to persist and ripen.... The process is thus performed: when the male palm is in flower, they at once cut off the spathe on which the flower is, just as it is, and shake the bloom with the flower and the dust over the fruit of the female, and, if this is done to it, it retains the fruit and does not shed it. In the case both of the fig and the date, it appears that the "male" renders aid to the "female"—for the fruit-bearing tree is called female.*

Theophrastus and his contemporaries did not realize that the bees they observed scrabbling around the plants were graciously introducing male "aid" to the female flower.

Two millennia later, the sexual nature of plants and their liaisons with bees were still only hazily understood. In 1682 Nehemiah Grew speculated that stamens were male plant parts and served somehow in the generation of seeds. Grew noticed the bee's interest in the male organs, although he did not connect it to the sex life of the plant. The stamens, he thought, were more like cafeterias. "Food for a vast number of little animals, who have their peculiar provisions stored up in these (stamens) of flowers: each flower becoming their lodging and their dining room, both in one."

Grew's contemporary, a German named Rudolph Jacob Camerarius, conducted a number of experiments on flowers in which he removed their masculinity by plucking the stamens. With the males thus "castrated," he found that female plants of nearby species failed to produce viable seeds. Scientists all over the world followed suit, brutalizing tulips, corn, beans, peas, and cabbages in search of their sexual secrets. Forty years of experiment later, the great Swedish botanist Carl Linnaeus catalogued the 7,700 species of plants then known in the world, classified and organized according to their arrangements of male and female parts. By the 1750s, the sexuality of plants was an established fact, though their methods of intercourse were still unclear. In the *Systema Naturae*, Linnaeus wrote of a special "messenger of love" needed to fertilize the flowers but was not specific as to who or what this emissary might be.

By the middle of the eighteenth century, male and female plant parts were at last being linked with their messengers of love, the bees. Arthur Dobbs, a backyard beekeeper and amateur scientist, observed honeybees in his garden at Castle Dobbs in Ireland. In a 1750 letter to a member of the Royal Society in London, he wrote:

> *I have frequently follow'd a bee loading the farina, bee-bread . . .*
> *upon its legs, through a part of the great field in flower; . . . Each load*
> *upon the legs of a bee is of one uniform colour throughout, as a*
> *light red, an orange, a yellow, a white, or a green, and is not upon*
> *different parts of the load of a different colour; . . . the presump-*
> *tion is, that it is gather'd from one species.*

Linking the pollen collection of the bees and the coition of flowers, Dobbs exposed an affair that had been a secret since the time of Aristotle.

*Now if the facts are so, and my observations true, I think that
Providence has appointed the bee to be very instrumental in promoting the increase of vegetables; . . . and at the same time they
contribute to the health and life of their own species. From the late
improvements made by glasses, and experiments made, in observing the Works of Nature, it is almost demonstrable, that the farina
upon the apices of flowers is the male seed; which entering the pis-
tillum or matrix in the flower, impregnates the Ovum, and makes
it prolific.*

Joseph Gottlieb Kölreuter wisely included the rest of the insect
world in the sex life of plants. In 1766 the German scientist wrote:

*In flowers in which pollination is not produced by immediate con-
tact in the ordinary way, insects as a rule are the general agents
employed to effect it, and consequently bring about the fertilization also; and it is probable that they render this important service
if not to a majority of plants at least to a very large part of them,
for all the flowers of which we are speaking here have something
in them which is agreeable to insects, and it is not easy to find one
such flower that has not a number of insects busy around it.*

Dobbs, Kölreuter, and other investigators had at last linked plants
and insects in an ancient and eternal sexual liaison. A few years
later, the German botanist Christian Sprengel made an important
discovery about how the plants chose their partners. Convinced as
he was that "the wise Creator of nature has brought forth not even
a single hair without some particular design," Sprengel experimented
with various flowers and determined that even if they had both

male and female parts (which 80 percent of them do), most were specifically designed to attract and welcome pollen-bearing messengers from other blooms. They "abhorred" their own genetic materials and endeavored not to produce weak offspring by self-fertilizing or cloning themselves. Sprengel's *The Newly Revealed Secret of Nature in the Structure and Fertilization of Flowers*, published in 1793, described the floral adaptations of over 500 species, all of which were designed to encourage insect visitation, cross-pollination, and stronger offspring. "Nature," wrote Sprengel, "seems unwilling that any flower should be fertilized by its own pollen."

In the religious and political climate of the late 1700s, the notions that plants were sexual, cross-pollinated, and enlisted the aid of insects met with considerable skepticism. The advances of Dobbs, Kölreuter, Sprengel, and other innovators languished until Charles Darwin appeared half a century later with exhaustive research that validated the earlier theories about plants, pollen, and the evolution of species. In 1860, three years after the revolutionary *On the Origin of Species by Means of Natural Selection*, Darwin published *The Various Contrivances by which British and Foreign Orchids are Fertilised by Insects*. Fourteen years later he wrote *The Effects of Cross and Self-Fertilisation in the Vegetable Kingdom*. In these volumes, Darwin verified that flower and vegetable species, like humans and animals, competed for and swapped different genetic materials to survive and flourish.

Over the eons since those first waspy bees, flowers have evolved not to please the poetic eye or enrich the florist industry but to appeal to insect pollinators and survive. If it were easy to attract bees and other visitors, all flowers would probably look the same. Instead, 250,000 known species of plants must compete for the pollinating attentions of 750,000 types of insects. Flowers have adapted their scent, appearance, architecture, and advertising in order to seduce insects, birds, and wind to capture the genetic material they transport. Plants that desire and require pollination from dung beetles, for example, have evolved darker colors and a lovely manure-like bouquet to attract them, as have some flowers that rely on flies. Red blossoms have become more vibrantly colorful in order to seduce the hummingbirds and butterflies that are partial to crimson. Blooms with deeper nectaries, like snapdragons and some clovers, rely on the longer tongues of bumblebees for pollination. Flowers seeking honeybees compete in a variety of ways, producing tastier nectar, brighter colors, wider landing areas, or more elaborate, timely, and appealing presentations of pollen. The evolvement of the natural world is a competitive beauty pageant, and bees, because of their particular appetite for pollen, are often the judges.

Bees are sought-after bachelorettes in the evolutionary dating game. Other pollinators—birds, bats, wind, and other insects—are also at work servicing their own specific clienteles, but none so effectively and universally as the bees. Ants, for example, also forage on the sweet watermelon nectar, but one plant at a time and without bodily fur or collection baskets, which limits a timely or copious sexual barter among several plants. Bats and moths forage at night, when most blossoms have closed. Butterflies, those bright flying swatches of color, are interested only in nectar and show no attraction or allegiance to pollen, as the bee does on each foraging sortie from the hive. Butterfly, bird, and wind pollination are wanton and scattershot compared to the bullet-like directness and determination of the bees. Their voracious and singular devotion to a particular type of pollen on each excursion makes bees perfect matchmakers between flowers that otherwise would have great difficulty meeting each other, let alone exchanging sexual materials in the time and quantities required.

Because of bees' starring role in the drama of pollination, we humans are indebted to them, directly and indirectly, for a third of our food supply. Visiting bees are required for the commercial production of more than a hundred of our most important crops, including alfalfa, garlic, apples, broccoli, Brussels sprouts, citrus, melons, onions, almonds, turnips, parsley, sunflower, cranberries, and clover. They are also beneficial, though not crucial, to fifty other crops, including asparagus, coffee, anise, endive, pears, apricots, peppers, coconuts, strawberries, and tomatoes. Bees bring us our morning coffee and, by pollinating the alfalfa crop that produces hay and then meat and milk, deliver the cream we lace it with. As E. O. Wilson, the acclaimed entomologist and honeybee fan, describes it, "one of every three mouthfuls we eat, and of the beverages we drink, are delivered to us roundabout by a volant bestiary of pollinators." We are also indebted to this bestiary for the evolution of the natural world around us. Pollination, as illuminated by Darwin, drives the plant world, and thus the humans and other animals that live in it. A third of our diet, and the very appearance and future of our planet, is the burden and promise that the bees carry on their furry bodies and legs.

At the beginning of the twentieth century, when the relationship between plants and pollinators was finally understood, it began to be exploited. Cultivators who accepted Darwin's ideas of evolution and pollination realized that the more bees they involved, the more "evolved" and fruitful those crops would be. This revelation coincided with late-nineteenth-century improvements in beekeeping that made bees more manageable, transportable, and applicable to agriculture. In 1918, John Harvey Lovell observed in *The Flower and the Bee: Plant Life and Pollination*:

> *When the cranberry bogs of Cape Cod and New Jersey bloom there are hundreds of level acres, which are literally covered with myriad pinkish-white blossoms.... On one side of a cranberry bog at Halifax, containing 126 acres, three or four colonies of bees were placed. This number was evidently inadequate to cover the whole field, and it was very noticeable that the crop of berries was largest near the hives, and became thinner and thinner as the distance from them increased. A small piece of bog entirely screened from insects produced very little fruit.*

According to Lovell, by 1918 honeybees were being applied to crops on an industrial level. "In Massachusetts, cucumbers are very extensively raised for market in greenhouses, and there are some 120 persons engaged in this industry, making use annually of more than 2,000 colonies of bees. Without bees or hand-pollination not a cucumber would be produced." Of apple pollination in New Jersey, he writes:

> *With the planting of orchards by the square mile their [wild bees'] number became wholly inadequate to pollinate efficiently this vast expanse of bloom. This difficulty is met by the introduction of colonies of the domestic bee. No other insect is so well adapted for this purpose. In numbers, diligence, perception and apparatus for carrying pollen it has no equal. In orchard after orchard the establishment of apiaries has been followed by an astonishing gain in*

*the fruit-crop; and today it is generally admitted that honey bees
and fruit culture must go together.*

Emphasizing the lucrative codependence of bees and agriculture, Lovell quotes one of the largest fruit growers in New Jersey as saying, "I could not do without bees. I would as soon think of managing this orchard without a single spray-pump as without bees."

Today in California, most almond growers would say the same thing. The $800 million business might be possible without spray pumps, but not without bees. Almond trees cover close to a half-million acres in the San Joaquin and Sacramento valleys, and the billion pounds of nuts they produce every year—three quarters of the world's supply—is reliant on the pollinating efforts of honeybees.

When the almond trees blossom in February, their branches form a vaulted white roof over acres of orchard. Thousands of colonies of bees are arranged in clusters at night beneath this delectable canopy. In a feat of timing and transport, and to ensure that the bees don't lose interest, they are parked only after the first of the blossoms have opened and begun their scented siren song. Many almond and fruit farmers mow down the area beneath and around the trees, so that not even a single dandelion can distract the bees from their purpose.

In the warm sunlight of morning, when the blossoms open, the bees leave their boxes like racehorses bolting from the starting gate. The orchard is soon humming and clouded with eager bees, hundreds inundating every tree. They hurry to the inch-wide pink blossoms, which have ten to thirty stamens waving above two plump ovules. Immediately, they begin their clambering search for pollen, dropping, knocking, and smearing it from their bodies onto the delicate, desirous stigmas. When one of the pair of ovules succeeds in being fertilized, the seed produced is an almond.

Each blossom not fertilized is one fewer almond, so California growers require a tsunami of bees, at least two colonies per acre. A swarm of insects jostles around every tree, feeding and fertilizing as it goes. But the triumph of almonds is not as easy as simply flooding an orchard with bees. Their flowers are hermaphroditic, but they are self-incompatible, which means they reject their own genetic tissue. The ovules must be plied with pollen from a different almond variety to produce fertilized seeds, so growers vary their rows of crop trees with plantings of slightly different sorts. As they gather and distribute almond pollen, bees must also take it from one type of tree to another in order to do the job for which they have been hired successfully.

California almond growers import more than a million hives annually to assist their genetically finicky crop. With their cold winters (almond trees require chilling during dormancy) and hot, dry summers, the valleys can't support this deluge of bees year-round. Bees hired for pollination are seasonal workers, billions of freelancers trucked to California each winter. In the largest pollination management program in the world, bees come from thirty-eight states to escape cold, nectarless winters and promote sex and almonds in northern California.

Of the three million commercial bee colonies in the United States, more than two thirds travel for pollination every year. Truck convoys loaded with screened-in towers of hives can be seen wending their way into the valley each winter. (And occasional swarming mishaps can be read of in the papers.) Like any traveling talent show, the business is run by an assortment of managers, subcontractors, and road crews, who make sure the performers show up on the appointed night in the right venue. Many beekeepers migrate with their livestock, especially those who don't rely on the brokers and managers who act as agents, arranging insect talent, contracts, payment, transportation, and fees with the growers.

After the almond blossoms wither, bees are relocated north to apples, sometimes to cherries and pears. In Florida they move from citrus to watermelon. Nationwide, rental rates per hive are between thirty-five and fifty-five dollars, so if beekeepers get their bees onto several crops, they can usually make more on pollination contracts than they can harvesting honey full-time. If the bees make honey while on pollination duty, it is a secondary, almost incidental product. Sometimes, when honey prices spike up enough, many keepers switch their bees to that liquid work, creating shortages of pollinators for the crop farmers. In a landscape vying for their attentions, there aren't always enough bees to go around.

The benefits of this pollinating partnership are almost immeasurable. A Cornell University study concluded in 1999 that the direct value of increased crop yields brought about by bee pollination was about $15 billion, while other studies indicate even more. Another researcher wrote, "If we consider the bees' value to be based on the fruit, vegetables, and seed resulting from pollination, we have a value that is about 150 times the value of honey and beeswax."

Unfortunately, the bee's alliance with modern farming has compromised her health. In the course of their agricultural good works and extended travels, bees are frequently exposed to pesticides sprayed in fields adjacent to those in which they forage, on roadside vegetation, or in contaminated pools of water. Anyplace bees hunt for food, they run the risk of contact with pesticides and other poisons, which can kill a colony in a matter of days. The widespread use of chemicals in farming and the decreasing amount of safe natural habitat for the bees make it increasingly difficult for them to escape this plague, and thousands of colonies are poisoned every year. Smiley has been relatively lucky, losing partial populations but never whole colonies. "Last year I took some bees up to cotton, and they sprayed the whole dang field," he laments. "I know I can't tell them when to spray their crops, but they could tell me. I had dead bees all over the place."

Science and agribusiness are struggling to save bees from poisonous fates by exploring pesticide alternatives, such as the genetic modification of crop plants. By splicing the genes of a particular bacteria into corn, for example, scientists have produced plants that are toxic to the European corn borer, a ruinous pest. This technique has been shown to increase crop yields and reduce the need for pesticides, and the corn product has been declared entirely safe for human consumption. Unfortunately, it has not yet been proven entirely safe for insects. A Cornell University study showed that pollen granules from genetically modified corn were toxic and fatal to the monarch butterfly caterpillars that fed upon them. If modern farming industries don't figure out how to differentiate good pests from bad, they will inadvertently endanger one of their most important business partners, the honeybee.

Urban development is another problem for beekeepers, bees, and people who enjoy honey and the benefits of cross-pollination (that would be all of us). The natural foraging grounds of bees are being paved and sodded over, and populations of wild and even commercially kept bees are threatened by the loss of habitat. Herbicides used to kill weeds (such as clover and gallberry) in gardens and on roadsides also deprive the bees of valuable food sources. Even in the relative wilds of Wewahitchka, the forage is changing fast. "More and more people are moving in," says Smiley. "That'll be a problem for me down the road. I've already lost two locations because of a housing development. In the next three to five years, I'm gonna lose more. Next thing you know we'll have a Super Wal-Mart here, and my bees will suffer." Super Wal-Marts, housing developments, and the like are imperiling the stomping and pollinating grounds of our beloved, necessary bee.

Bees are also defeated by ignorance and indifference. Recently I was watching television in Connecticut while bees were playing and pollinating outside the house. Flipping channels, I came upon a reality show in which a couple, with great horror and disgust, discovered honeybees in their backyard garden. It was as if an army of rats had set up camp in their ivy, and the couple immediately went to Home Depot to purchase a can of Raid. The husband was later shown as conquering hero, spraying the bees to death. I turned off the television, hoping my bees hadn't seen the violence through the windows and wishing people would learn their birds and bees. Not long after that a woman told me she hated bees because her husband had been stung by a wasp.

At the end of June, when the watermelon vines have ceased flowering, Smiley returns to pick up his pollen-sated workers. He comes late in the afternoon, when the bees are at rest, and loads them back on his truck. Some Mexican pickers are already at work harvesting the early melons, keeping a wary eye on the beehives, unaware that the inhabitants are interested only in nectar and pollen. The prized granules are gone now, but all parties to the pollination have reaped rewards. Smiley has his rental fee and several hundred pounds of watermelon honey. The bees have supplied their young with protein, the watermelon flowers have socialized with members of the opposite sex, and their farmer has a plentiful crop of rounded melons.

Smiley arrives home to the brightest-colored house in Wewahitchka. Amidst the beige and graying homes on Bozeman Circle and the dark green and brown towering woods beyond, his future abode is an emphatic flamingo pink. The stucco man came in the beginning of June and coated the house in a plastic veneer that is a good deal more colorful than Smiley or Paula had intended. Paula had envisioned an understated southern pastel, but they ended up with a bright Caribbean coral that is more Miami than Wewa. "It's quite a bit pinker than what she had originally planned," says Smiley. "But it's pretty. We'll live with it." The new house is a one-story bungalow with white trim, a gray roof, and elegant arches over the entry and garage bays. It looks a little out of place in the neighborhood, but in a cheerful, confident way. Paula, who loves the beach, declares that they have brought it to Wewa.

FIVE

The Sting

The Bee and Jupiter

*A queen bee from Hymettus flew up to Olympus with some fresh
honey from the hive as a present to Jupiter, who was so pleased
with the gift that he promised to give her anything she liked to ask
for. She said she would be very grateful if he would give stings to
the bees, to kill people who robbed them of their honey. Jupiter was
greatly displeased with this request, for he loved mankind: but he
had given his word, so he said that stings they should have. The
stings he gave them, however, were of such a kind that whenever
a bee stings a man the sting is left in the wound and the bee dies.*

Aesop's Fables

*He is not worthy of the honeycomb,
that shuns the hives because the bees have stings.*

William Shakespeare

Float like a butterfly, sting like a bee.

Muhammad Ali

One morning in June, Smiley walks, coffee cup in hand, from his old beige house to the new pink one to check on the progress of the sheetrock hanger. He's had to monitor the house work pretty constantly, which irritates and frustrates him and takes him away from his bees. For the sheetrock, for example, he had to interview four different guys for the job, and he's still not sure about his choice. "One guy gave me a quote and my jaw about dropped to my knees," he recalls. "I said, 'Listen, I just want to live in it' and sent him on his way." Smiley worked in construction long enough himself to know that the hanger he finally chose is reasonably priced and good, but the whole project is taking longer than he would like, and it's making him restless and grumpy. "The workers here always need something," he complains. "And the bees always need something. I'll be on my way to a beeyard, and I'll get a call about a problem and have to turn around and come back. Seems like I need to be here all the time, right when my bees need me."

He is relieved and uplifted when his bee help arrives and he can go to work doing something he really enjoys: robbing honey in a sting-filled yard. Smiley, George, and Keith ride over to a yard hidden in the woods just south of town. Approaching on Route 71, Smiley slows the truck while he scans the roadside for the path that leads to the yard. There it is: a barely noticeable narrow track through an area of gallberry bushes and low young pine trees. As he turns off the road and into the brush, fat bugs dive at the truck and glance off with loud metallic pings. Pine branches swish past as if in a car wash. Sun-warmed snakes whisper from the track into the tall grass. The air here hangs so thick with humidity, dust, and pollen that he can see it. When he parks the truck and opens the door to the day's heat, a chorus of cicadas and crickets builds to a rattling roar, then ebbs into silence.

The hives in this yard have been here since the end of the tupelo harvest, making gallberry honey for almost a month. The crew veils up, smokes up, and starts robbing. Soon three men, a truck, and three million bees are all bumping into one another in the small hot clearing. Smiley thinks maybe the bees were oversmoked by a nervous Keith, or maybe it's too hot or the nectar's running out, but the bees are up in arms. Bombing, taunting, and strafing, they valiantly try to protect their honey from the invaders. In straight menacing patrols they fly low, determined, and fast across the yard, buzzing furious and loud. Bees dive onto and bounce off their hats like hailstones, and the sting count is high. Yelps and muttered curses alternate with the roar of the blower. George gets stung under his arms, swears, and does a violent chicken dance across the yard holding an eighty-pound box of honey.

Generally, bees are pacifists; they would rather flee than fight. They sting to protect their honey, their young, and their queen, and only when provoked. The invasions of the harvest can constitute extreme threat and provocation. Bees get understandably ornery after weeks of disruption, dislocation, butyric acid, smoke, and the repeated theft of their food. Often during the harvest, it seems as if the bees are at war with their keeper. When he gets stung, Smiley winces stoically and tries to reason with the attackers. Sometimes they don't listen. Occasionally, when the resistance gets too uncomfortable, Smiley and his team withdraw to the side of the yard for a few minutes and wait for the bees to calm down. They drink water, chew tobacco, or smoke cigarettes between rounds in the pugilistic yard.

When he takes a break, Smiley returns to the truck through clouds of indignant bees. He swigs water and looks down at his forearms, which appear to have been run through a patch of burrs. "I hate it when they fight me," he sighs, taking off his hat and veil to look more closely at his wounds. Amidst the hairs and freckles on his arms is a sprinkle of what look like brown and fuzzy sesame seeds. These are the venom sacs of the bees that stung him in the last hour or two. Their stingers are below the sesame seeds, embedded in his skin. The perpetrators have gone, but they leave their weapons behind, miniature poison darts studding his arm. Some of the venom bulbs are still pumping faintly.

Using the back of a longer fingernail or sometimes the end of his hive tool, Smiley nudges it beneath each little globe, then pushes, sliding the barb gently yet quickly out of his skin as delicately and decisively as if he were removing a splinter. When he has pushed the stinger free, he flicks it away with his fingernail as if shooting a marble. In his earliest beekeeping days, he grabbed the whole sac, which simply squeezed in more venom, so he's perfected this scrape, push, and flick technique. What comes of each scraping, before the flick, is a shiny dark thorn, as if from a miniature rosebush, attached to the furry sesame seed sac.

Normally this weapon is retracted within the bee's lower body but poised, like the landing gear on a plane, to drop and lock into place when she gets annoyed. As alarm, excitement, and pheromones motivate the bee, she lowers her sword into position at the end of her abdomen. It locks into position at a right angle, like the needle of a record player. At the moment of the bee's greatest distress, muscular abdominal plates punch the stinger into flesh.

The bee's sword is composed of three parts. On top is the stylet, a grooved bulkhead that starts at the venom bulb in the abdomen of the bee and tapers to the scalpel-sharp stinger point. Flanking the stylet are two retractable rods, called lancets, which are serrated like steak knives along their outside edges. Under a microscope,

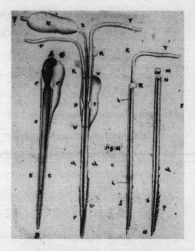

Jan Swammerdam's interpretation, using an early microscope, of the bee's stinging mechanism.

the stinger looks more like the snout of a swordfish, smooth and flat on top and jaggedly barbed on both sides. When the bee stings, the lancets scissor and grab their way into the flesh, clearing an alley through which the pointed shaft of the stylet can easily glide into the wound.

The three surfaces form a ratcheted needle through which venom can flow into the victim, abetted by the muscular pumping of the abdomen. The barbs of the stinger lodge it in the flesh like a fishhook, maximizing the time of poison delivery. When the fighter struggles to fly away, her weapon remains embedded, with bits of abdominal tissue and hair from the bee's body giving the abandoned sac its slightly furry appearance. For the bee, the sting is a dismemberment too great to survive. She will fly away and die within seconds, even as her venom sac is still valiantly pumping poison at the sting site. This kamikaze act occurs only when the target is a human or similarly fleshy mammal. The bee's weapon doesn't embed in other insects (which might try to invade the hive to steal honey), on which it can be used repeatedly if necessary. Only female worker bees are asked to make the final defensive sacrifice against threatening mammals: The queen bee will use her smaller, barbless sword only against other royalty, and drones are not issued any weapons at all.

A bee sting is an incomparably sudden, painful shock, like a car door slamming on fingers. When the surprise fades, the sensation shifts to that of glowing hot coals under the skin. Some people who have been stung compare it to shards of slicing hot glass. Maurice Maeterlinck, in his *Life of the Bee*, put the pain more poetically: "Her sting, which produces a pain so characteristic that one knows not where with to compare it; a kind of destroying dryness, a flame of the desert, rushing over the wounded limb, as though these daughters of the sun had distilled a dazzling poison from their father's angry rays."

The stinging mechanism takes thirty to sixty seconds to unload its dazzling poison. Up to half the dose is comprised of a toxic protein called melittin, which bursts blood vessels and damages tissue. Melittin's purpose is, quite simply, to hurt. A hint of it also encourages the body to release histamines, which defend against the venom and in the process produce itching, redness, and dramatic swelling. At first, the sting site is a hot red pinprick. As the melittin and histamines go to work, the flesh around the wound warms and rises to a painful burning welt the size and thickness of a nickel. In the first hour or so after the sting, the nickel blossoms into a quarter and then a puffy pancake. Over the next several hours the blushing pain and uncomfortable swelling can creep as far as a foot from the sting site. The results are dramatic—afflicted limbs double in size, nearby joints submerge in swollen flesh. Stung hands become clumsy catchers' mitts or inflated surgical gloves, and attacked ankles no longer bend or fit into shoes. The creeping swell is accompanied by aches and feverish chills as the body attempts to cope with the toxins and histamines. All of this awkward discomfort is caused by about 100 micrograms of poison, or the equivalent of a grain of salt. The surprising trauma of bee stings is not from the puncture itself, which is a clean, minute slicing of the skin, but from the unleashing of the flesh-altering venom and the storm of histamines that follows.

A mischievous team of enzymes called phospholipase A_2 and hyaluronidase act as a sidekick to the mellitin, aiding its painful rampage. They have an emollient effect, like water on a dry sponge, on the acids that hold tissues together, allowing the toxins greater, freer, and more painful spread into the flesh. Bee venom also delivers small amounts, less than one percent each, of dopamine and norepinephrine, neurotransmitters that accentuate fear and excitement and make the sting unforgettable. The final elegant efficiency of the sting is the alarm pheromones it releases. As she is embedding her poisonous dart, the bee also emits a scent that calls reinforcements to the sting site, increasing both the victim's dread and his potential dose of poison.

Day two of a bee sting is similar to a sprain—burning pain, heat, clumsiness, and freakish swelling. Depending on how much venom is injected and how a body responds to it, the effects can last for several uncomfortable days. While many of the eighteen active components in each dose are not yet entirely understood by scientists, they clearly conspire in a sophisticated way to produce, expand, and protract the victim's pain and discomfort. Stings are not intended to kill or cause any permanent damage, other than fear.

Most people experience a dramatic yet local reaction to bee stings. Even if a stung arm swells to the size of a bolster pillow and nearby fingers threaten to explode, this is still considered a "local" reaction, because it is related to the venom at the sting site. A very few people experience a systemic allergic reaction to the venom, which might cause a rash or hives anywhere on the body, as well as dizziness and vomiting. The most extreme and rare allergic reaction, anaphylaxis, causes the throat to swell shut and blood pressure to drop, leading very quickly to shock and unconsciousness that is fatal if not treated immediately with a powerful antihistamine. Although it attracts a lot of terrifying press and hearsay, anaphylactic shock from bee stings is extremely rare. Just in case, beekeepers often keep epinephrine on hand if a new visitor to their hives happens to be in the 1 percent of the population that is hypersensitive to bee venom.

Insect stings are responsible for 50 to 100 deaths in the United States each year, and only half of those fatalities come from honeybees. During the same period, more than 4,500 people die from drowning, 1,600 people are killed due to firearms accidents, and more than 400 deaths result from "slips and falls while walking." For the average person, the chances of committing suicide, being struck by lightning, or falling down stairs are magnitudes greater than the possibility of being killed by a honeybee's venom. In spite of these odds, many people continue to fear bees (more than walking, apparently) as dangerous and deadly.

This fear is the bee's best defense against the robbers of the world. In *Langstroth on the Hive and the Honey Bee*, the author wrote, "The sting of the bee, a terror to so many, is indispensable to her preservation. Without it, the attraction, which honey represents to man and animals, must have caused the complete destruction of this precious insect, years ago." During the harvest, bees can be seen and heard flaunting their dangerous reputation in an attempt to save their coveted honey. Even after a calming blast of smoke, the pitch of the bees' buzzing is a heightened roaring whine. When the lid comes off, several inhabitants will fly straight out of the hive, trying to bomb and intimidate the intruder. They'll dart around angrily, diving, nudging, bumping, and buzzing in a showy attempt to cow and fluster. Most passersby or newcomers to the hive are convinced by this act. As they back away or run flailing, the bees relax and remove their war paint.

If and when these blustery tactics fail, the bee unleashes weapon number two, her stinger, and takes aim on her suicide mission. Armed and under duress, bees will dive for the nearest patch of unprotected flesh, attracted to its warmth and odor. Places where the skin is thin and tender—on the hands, neck, and joints—are even more effective and appealing. Contrasts also interest them, which is why Smiley tries to remember to take off his wristwatch, a black plastic digital Timex. If he ventured into an aggressive hive with it still on, the bees would go directly for the dark band, giving him a red memento bracelet of stings around his wrist. Hat and veil protect the most delicate combinations of warmth, odor, and contrast—nostrils, eyes, mouths, and the shadowy curled interiors of ears. Smiley thinks it hurts the most to get stung on the nose. "Mouth," says one assistant. "Ears" is another's final answer. Desiring not to explore this debate too much, most experienced beekeepers wear veils during the harvest.

A first sting (or set of them) is usually painful and frightening but otherwise uneventful because the body is not yet sensitized to bee venom. On subsequent sting episodes, the venom is recognized and combated, and the allergic reaction is impressively more dramatic. After many, many doses, the melittin loses its disruptive power. The body builds up a tolerance, releases fewer histamines, and also produces cortisol, a steroid that can act as a powerful natural anti-inflammatory. Smiley, who has absorbed quantities of venom, is now virtually immune to the poison but not the pain. "I still feel it, believe me. I've just learned to ignore it," he says. At the end of an adversarial day in the beeyard, Smiley plucks stingers out of his arm as easily and quickly as if they were hairs off a sweater. In the place of each bite he will have a tiny red welt, more like a freckle, that will subside and disappear in a matter of hours. For him, stings are a mere nuisance, a mild annoyance, like a mosquito bite. Indeed, panhandle mosquitoes are so big, plentiful, and aggressive that sometimes they are a greater nuisance than agitated bees.

It takes a violent allergic reaction to bee venom, or an extremely large amount of it, to kill an adult human being. Africanized bees have earned the ominous nickname "killer" bees because of their ability and preference, as a group, to deliver massive amounts of poison. The number of bee-related deaths is likely to increase in the next few years as Africanized bees continue to arrive and spread throughout the United States. Their invasion is a story of well-intended bee science gone badly wrong. The experiment began in southern Brazil in the 1950s, when honeybees, not native to South America, were imported from Europe in hopes of jump-starting a honey industry. Mild-mannered European bees, however, did not respond well to the sultry Brazilian climate, and their adaptation and production were disappointing. The Brazilian government sought to import more suitable, heat-friendly workers and took shipments from South Africa. These African bees, raised in a much more expansive, hostile landscape, were lean, mean, and nomadic, physically and temperamentally different from their docile European cousins. African bees, like packs of wild dogs, are fierce, dominant, and aggressive in every way that the Europeans, like groomed, shiny golden retrievers, are gentle and submissive.

A year after they arrived in Brazil, a gang of African queens and their posses escaped into the jungle and began overturning weakened European bee thrones. Within three years, the race had subjugated, usurped, and Africanized the bee nation of Brazil. Since then, the Africanized bee has progressed northward at a phenomenal pace, almost 200 miles a year, leaving a trail of Africanized colonies and demoralized beekeepers in its wake. Venezuela was conquered in 1973, Mexico in 1986, and in 1990 killer bees arrived in the United States, in southern Texas. From there, they have spread and adapted north and west, through the warm, dry, South Africa–like states of New Mexico, Arizona, Nevada, and California. The first California discovery was in October 1994. One year later, more than 8,000 square miles were declared Africanized.

Africanized bees look only a little different from the gentle European strains that have been in North America since the seventeenth century. Their physique is the same, except that they are about 10 percent smaller and 25 percent lighter. They reproduce faster, change abodes frequently and have shorter life spans than their European cousins. The venom from the two creatures is virtually the same, so the killer difference is behavioral. Whereas European bees sting reluctantly and desist as soon as the perceived threat is gone, Africanized bees seem to relish the attack. They will go immediately, even preemptively, on the offensive, responding by the hundreds, pursuing intruders for hours and over great distances to deliver their deadly dose. In one Africanized bee death, reported in Costa Rica in 1986, a botany student who had stumbled across a hive was stung by eight thousand Africanized bees, an average of twenty stings per square inch of his body. Authorities say that the lethal dose of venom is 8.5 stings per pound of body weight.

Mark L. Winston, in his book *Killer Bees: The Africanized Honey Bee in America*, describes visiting a recently Africanized apiary in Suriname:

> *We parked about a half-kilometer from his beeyard, put on two layers of clothes under our bulky beesuit coveralls, and carefully secured our veils and gloves to leave the bees no room to enter. Then we lit the largest smokers I had ever seen, bellows-like instruments that burn burlap, old sheets, cardboard, dried cow patties, or whatever is available to generate smoke to pacify the bees. Only then did we approach his colonies, and I should have been warned by these elaborate precautions.*
>
> *Merely walking toward the colonies elicited a massive response on the part of the bees, so that the situation was out of control before we smoked and opened our first colony. Bees were everywhere, banging into our veils and helmets with such ferocity that we could barely hear each other and stinging through our layered clothing. It was a hot, humid day and the combination of sweat, noise, and stings forced us to retreat after examining only a few*

colonies. The bees followed us all the way back to the car, and we had to keep our equipment on until we were far out of their stinging range. As we drove off, we could see the farmers swatting at bees and two of their cows were being stung; we had to stop and move the animals farther away to safety.

Africanized bees do produce honey, but given their smaller colony size, mobility, and surliness, the harvest is unreliable and unpleasant. Some beekeepers, wearing lots of protection and a determined smile, persist in harvesting honey from Africanized beehives, but most honey farmers dread the looming threat of being Africanized. Beekeepers from Salt Lake City to Wewahitchka have been looking into their supers and over their shoulders, wondering who and where will be next. Wewahitchka, so far, is Africanized-free. "Maybe they're avoiding the mites and the rain around here," says Smiley. "I don't know. I don't know what its going to be like if they come here. But you can definitely work 'em, you just have to leave 'em alone. I talked to one guy in Texas who stacks a bunch of supers on the hive and then leaves them alone for the whole season." Texas-style management techniques might soon be in order in Florida, as no effective means of deterring the African arrival has been developed. "If they come here, I'm just gonna build me a screen cage to walk around in so I can watch George work," jokes Smiley as George moves boxes around the yard. Scientists are trying to breed the African scrappiness and aggression out of them, but they seem to be persistent, prevailing traits. As the Africanized honeybee continues to conquer the United States, we will all have to get used to a more warlike honeybee and the increased threat of deadly swarms. If and when these bees are fully established in North America, it is predicted that two hundred people a year could die from mass envenomation.

Swarms of killer bees are a special effect in apocalyptic science fiction, but death and injury due to venomous insects have been a real and terrifying part of human history. Terror associated with locusts and other plagues of flying, stinging insects is clear in the Bible. In the book of Exodus it is written: "And I will send hornets before you, which shall drive out the Hivite, the Canaanite, and Hittite before you." In Deuteronomy, stinging insects are heaven-sent: "The lord your god will send hornets among them, until those who are left and hide themselves before you are destroyed." Some of these biblical hornets were likely a form of early African bee, vicious, deadly, and feared. In the ancient world, from Mount Olympus to Valhalla, bees were deliverers of God's wrath. Their stings were loaded with divine power, sacred mystery, and, of course, punishing, ominous pain.

In a superstitious world of sharpened sticks, hot oil, and sling-shot stones, bees were impressively high-tech weapons. They could be persuaded to chase and attack, hurt, cause frightening disfigure-ment, and inspire unparalleled fear. Warriors learned early to exploit the protective instincts of the bees, transforming them into power-ful weapons. From ancient Greco-Roman battles into the twenti-eth century, man has taken advantage of these flesh-seeking missiles to inflict pain, punishment, and torture.

Greco-Roman warfare often included the strategy of tunneling toward enemy fortifications. Aeneas Tacticus, the world's first mili-tary historian, noted in the fourth century B.C. that bees and wasps could be released into these tunnels to plague the advancing bur-rowers. Three hundred years later, during the Mithridatic Wars in the Black Sea region, defenders against the Roman general Lucullus cut openings into his tunnels from above and "thrust bears and other wild animals and swarms of bees into them."

Before sophisticated box hives were invented, bees kept in twig, straw, or clay vessels of various sizes were adapted as weapons. Bee grenades could be hurled through the air or dropped on enemies. By the time they reached their target, the projected bees were outraged, ready to explode in a fury of stings. In Book III of his *History*, Herodian describes the second-century siege of Hatra, in modern-day Iraq, and the use of bees as weapons:

> *Every kind of siege engine was used against the walls (of the city) and no technique of siege operation was left untried. But the people of Hatra rigorously defended themselves by firing down missiles and stones onto the army of Severus below and causing them a good deal of damage. They made clay containers filled with little flying insects that had poisonous stings, which were then fired off. When the missiles fell on to Severus' army, the insects crawled into the eyes and exposed parts of the skin of the soldiers and stung them, causing severe injuries.*

For naval battles, the Romans developed special shipboard swarm catapults. They raised bees and kept them in lightweight, fragile earthen hives for the sole purpose of lobbing them onto enemy ships. Angry bees would so unnerve the opposing sailors that they often jumped overboard to escape. Turkish and Spanish pirates in the Mediterranean are said to have used this technique to capture other vessels.

In the Yucatán peninsula, the Quiche, one of the Mayan tribes, were also using bees as weapons. A tale in the Popul Vuh, the creation story of the Mayans, tells of a besieged Quiche hillside village. The tribesmen craftily collected hornets, wasps, and bees in pots and gourds and placed them strategically around the town. When the enemy advanced, the Quiche hid, allowing them to enter.

> And then the gourds were opened up, and the yellow jackets and wasps and bees were like a cloud of smoke when they poured out of the gourds. And the warriors were done in, with the insects landing on their eyes and landing on their noses, their mouths, their legs, their arms. There were insects going after every single person.... No longer able to hold onto their weapons and shields, they were doubling over and falling to the ground, stumbling. They fell down the mountainside. And now they couldn't feel a thing when they were hit with arrows and cut with axes.

Medieval warfare relied heavily on fear and handy materials, so bees were preferred projectiles, readily available in most villages and castles. (Bags of snakes and the dead bodies of plague victims were two other popular catapult choices.) Being lightweight, portable, and fragile, beehives could be handily dropped over the ramparts onto enemy troops as they approached. There are dozens of examples of these bee bombs from the tenth to the seventeenth century, when better weapons came along. In the eleventh century, Henry I's army tossed hives at the Duke of Lorraine's troops. During the first crusade, at the siege of Maara, the Muslim inhabitants made defensive use of "stones, beehives swarming with bees, fire, and even quicklime," while desperately trying to keep the Christian enemy from the gates. During the Thirty Years' War, in the 1600s, the residents of Kissingen in Bavaria hurled beehives at invading Swedish

Beehives thrown over castle walls were a common and persuasive defense.

cavalry. Horses are perhaps more spooked by bees than are men, so hives tossed onto cavalry were a choice deterrent.

A manuscript from this period includes a diptych drawing of a device used for launching bees. It appears to be a combination of a windmill and a slingshot, with the paddles of the mill equipped with nets to hold hives of bees. A helmeted and mailed worker loads skeps into the nets as the paddles rotate. The second panel shows hives flying through the air towards a walled fortress whose flailing inhabitants are a portrait of distress.

In addition to military service, bees were historically used for personal protection. Average folks who had beehives in the backyard could set them against thieves and scoundrels the same way they had been turned against invading armies. In Ireland in the sixth century, a nun named Gobnat used her bees as if they were a can of modern-day mace. She was tending her beehives when thieves sneaked into the convent to rob the defenseless nuns. Gobnat shook the hives onto the raiders, causing them to flee in fear and pain. Convents did not have many weapons at their disposal, but they often had apiaries, and in several historical accounts agitated bees come to the aid of a shaken sister.

Before household safes or bank vaults, beehives were a convenient place for ordinary citizens to keep money and important items, as it was unlikely that thieves would think of or dare to search them. The poet Virgil is said to have stored his valuables in his backyard hives. Eva Crane, interviewing beekeeping families in Europe, found that many of them had kept treasured items in their hives until the middle of the twentieth century. Some probably still do.

Bees were also employed in matters of the heart and hearth. Edwin Teale, in his introduction to *The Life of the Bee*, wrote: "Peasant girls in central Europe used to lead their lovers past beehives, believing an old superstition that if a man who had been unfaithful should pass by, the insects would rush out and sting him." In France, cheating husbands avoided their backyard beehives, as it was known that bees would sting a man who had been unfaithful. All over Europe, bees were known to keep domestic peace, for they would not produce honey for or stay in a family that was quarrelsome, dysfunctional, or unhygienic. If the atmosphere didn't suit them, they would sting and leave. Charles Butler wrote of the proper atmosphere in the seventeenth century:

If thou wilt have the favour of thy Bees that they sting thee not, thou must avoid such things as offend them: thou must not be unchaste or uncleanly: for impurity and sluttishnesse (themselves being chaste and neat) they utterly abhore: thou must not come among them smelling of sweat, or having stinking breath, caused either through eating of Leekes, Onions, Garleeke, and the like; or by any other meanes: the noisomenesse whereof is corrected with a cup of Beere, and therefore it is not good to come among them before you have drunke: thou must not be given to surfeiting and drunkenesse: thou must not come puffing and blowing unto them, neither hastily stir among them, nor violently defend thy selfe when they seem to threaten thee.

As physical punishment, bees were cheap, effective, and globally employed. In 1625 Samuel Purchas published *Purchas his Pilgrimes: contayning a history of the world in sea voyages and lande travells by Englishmen and others*, in which stories of world adventure and oddity were compiled in four massive tomes. Purchas described a tribe in Africa, the Niellim, which used beehives to test for innocence in crimes. If the accused plunged his hand into an active beehive and was stung, he was guilty (and, conveniently, already well punished). Centuries later, in the Cameroon, bees were still at work administering justice. Carl Seyffert wrote, "bees are used as divine judgment, by which people who have sworn a false oath are killed." Military crimes were also punishable by bee sting. William of Malmesbury, a Roman Catholic monk and historian who lived from 1090 to 1143, reported that an army commander named Robert Fitzhubert "used to expose his prisoners, naked and rubbed with honey, to the burning heat of the sun, thereby exciting flies and other insects to sting them."

Honey has also been used as a seductive instrument of war. Plants of the rhododendron, azalea, and oleander families produce grayanotoxins, powerful poisons. Honey made from the nectar of toxic plants can produce a narcotic, sometimes deadly effect. In 1794, Benjamin Barton, a Scottish physician and botanist who lived in the United States, wrote that small amounts of some types of heather honey made him feel as if he had taken "a moderate dose of opium." Dizziness, numbness, psychedelic illusions, giddiness, and impaired speech are a few of the symptoms of toxic honey ingestion. These conditions can progress to delirium, vomiting, paralysis, and even death. In lands where certain plants were plentiful and the conditions were right, malignant honey could be slipped to the enemy, rendering them dazed or unconscious and easily vanquished. The Black Sea region of Turkey, for instance, was notorious in antiquity for its poisonous plants and for the "crazing honey" used in battle. In 386 B.C. the Greek historian Xenophon described in his *Anabasis* an incident involving Greek troops on their way home from a foreign campaign.

> *After accomplishing the ascent, the Greeks took up quarters in numerous villages, which contained provisions in abundance.... The swarms of bees found in the neighborhood were numerous, and the soldiers who ate of the honey all went off their heads, and suffered from vomiting and diarrhea, and not one of them could stand up ... Those who had eaten a great deal seemed like crazy, or in some cases dying men. So they lay there in great numbers as though the army had suffered a defeat.*

Strabo's *Geography* of the first century A.D. describes honey used similarly against the troops of the great Roman General Pompey. "The Heptacometae [the local people] cut down three maniples [regiments] of Pompey's army when they were passing through the mountainous country; for they mixed bowls of the crazing honey which is yielded by the tree-twigs, and placed them in the roads, and then, when soldiers drank the mixture and lost their senses, they attacked them and easily disposed of them."

Bees' appeal and power as weapons faded as that of guns and chemicals grew, but they continue to participate in modern warfare. Their most famous recent skirmish, in what is now Tanzania, is known as the Battle of the Bees. On November 4, 1914, as German and English troops faced off in the African bush, machine-gun bullets hit local hives, which were hollowed-out logs hung from tree boughs, and wrathful bees stung in force, causing both sides to retreat in agonized surprise. One royal soldier who remained at his post was reported to have sustained three hundred stings. For years afterward, it was claimed that the Germans had calculatedly used the hives as weapons, but they suffered the same injuries as the British. The bees seem to have been the only victors in the battle, as neither side wanted to go near the hives again.

In the Vietnam War, native fighters used bees deliberately against American troops. In their book, *The Tunnels of Cu Chi*, Tom Mangold and John Penycate interviewed Nguyen Chi, a guerrilla, on the ingenious use of bees in the jungles of Vietnam.

> We studied the habits of those bees very carefully, and trained them.
> They always have four sentries on duty, and if these are disturbed,
> or offended they call out the whole hive to attack whatever disturbs

them. So we set up some of these hives in the trees alongside the road leading from the post to our village. We covered them over with sticky paper from which strings led to a bamboo trap we set in the road. The next time an enemy patrol came, they disturbed the trap and the paper was torn from the hive. The bees attacked immediately: the troops ran like mad buffalo and started falling into our spiked punji traps. They left carrying and dragging their wounded.

There will undoubtedly be a place for bees in future wars. For the past decade, the American military has been testing their potential as special agents in the wars on drugs and terrorism. Bees are as sensitive to odor as dogs and can be trained to buzz in on drugs, explosives, land mines, and chemical weapons. It takes less than two hours to condition a hive of bees to reject flowers and, rewarded by sugar, seek contraband scent instead. Bees successfully locate explosive chemicals more than 99 percent of the time. They are more accurate than dogs, cheaper to train, feed, and maintain, and (sorry, dog supremacists) have perhaps a more ancient and illustrious history in the military and law enforcement.

Experience in war and crime prevention does not protect bees themselves from larceny, however. As pollination and honey prices rise, honey and bee theft are increasingly frequent agricultural crimes. Poachers sneak into yards in the dark of night to steal loaded honey supers. Sometimes they abduct entire colonies and put them to work making honey or pollinating crops for farmers who don't know that the bees are kidnapped. Smiley has had boxes stolen from him a couple of times, just a hive or two, but as he says, "You steal a few here and a few there, and before you know it you got a pile of bees." Beekeepers paint or brand their names onto their boxes and frames, but words aren't daunting to bee burglars. Recently, microchips, the kind used to track pets, have been installed in some commercial hives, allowing owners to locate stolen bees and equipment and more effectively deter trespassers. Signs affixed to these microchipped supers warn that the boxes are traceable and that thieves will be penalized—unfortunately, in cash fines and legal penalties instead of public stingings.

"People will go to an awful lot of trouble to get their hands on honey," says Smiley, shaking his head as he tells stories of various thefts. A colleague of his once had a yard robbed of all its honey supers overnight. "They knew exactly what they were doing," recounts Smiley. "They came in with a truck. Middle of the night. Took the covers off, swiped the honey supers, and left the bees in the yard." Nobody ever caught 'em." He hopes the bees defended themselves and their food aggressively.

Stings are a cost of doing business in honey, no matter what side of the law you are on. It's the price paid for the bees' sweet loot. A reasonable fee, Smiley thinks. He's dealing with an ancient, powerful, and noble fighting force. He respects the bees for doing their job, and he's got to do his. Apologizing to all the bees that perished in the morning skirmish, he finishes scraping tiny daggers from his arm, takes a gulp of fortifying coffee, pulls his veil back down over his face, and strides back into the fray to steal more honey.

ARROGANCE

My first season with bees, I was cautious, respectful, and pretty heavily armored every time I went into the hive. That was a sting-free year. By the end of my second beekeeping summer, I had grown cocky. One afternoon, as dusk was descending, I was in the house showering when from the bathroom window I spied the queen excluder leaning against the side of the hive. I had forgotten to replace it during a visit that morning. I could have left it there until the next day, as it was unlikely that the queen would go up into the honey super and litter my crop with eggs overnight, but I decided to sneak the excluder back into the hive before I went out to dinner. Freshly bathed, shampooed, and blow-dried, I marched up the hill wearing a purple tank dress with my hair hanging free to my shoulders.

I didn't bother lighting up the smoker, because I was delicately perfumed, in a hurry, and planning on being in the hive for less than thirty seconds. I figured the bees would hardly notice me discreetly slipping in with my excluder like an unobtrusive secretary dropping a file into a drawer. My gloves and veil were in the garage and too far away and grubby for me to put on. Besides, I was an expert.

Pulling the lid off the hive, I was greeted by the gentle murmur of bees finishing up their day's work. Leaning over the heavy honey super, I grabbed it by both sides and lifted. A posse of bees flew out to investigate the disruption and immediately got stuck in my hair. Dropping the box back down on the hive, I stood as still as a statue. Hanging my head down, I hoped the bees would extricate themselves from the hairy trap. They tried, unsuccessfully. With their furry bodies and legs dappled in honey from their work, they tangled in my thin brown hair like sticky lollipops, and every frustrated kick enmired them more. I could hear their distressed high-pitched whine and feel buzzing vibrations on separate parts of my head like little bursts of electricity. The statue crumbled, and I began jumping wildly up and down, running my fingers like a comb through my hair while maniacally shouting at the bees and myself to be calm. Not surprisingly, this made the bees more terrified and seemed to weave them deeper into my hair. I exhorted them to leave the trap while apologizing because I knew they couldn't and that they (or maybe I) would die. By now the bees were fighting me: terrified, furious, and instinctively aggressive. Apologizing one last time, I slapped my head hard several times as I danced, like a swimmer trying to knock water from her ears. I intended to crush the bees onto my scalp, killing them before they stung me. But bees are quick on the draw, and I deserved a lesson; I think each one gave me a dose of venom before she died.

My date and I spent part of the evening locating and extracting six stingers from my scalp. Each bull's-eye was surrounded by a welt the size of a mosquito bite. I was shaky, embarrassed, and, as it turns out, lucky. This was my first introduction to the bee's sting, so I had almost no allergic response to the venom. I might have had a swollen beekeeping head before I went into the hive, but I escaped receiving a potentially dangerous one from the bees. From that moment, I learned to approach the hive as a privileged exchange, never an afterthought or a chore.

I hate to admit it, but the bees caught me being cocky again. The second time I got stung was with Smiley, on the first day of my first visit to Wewahitchka. I was overwhelmed by the strangeness of the place, the panhandle heat, and the scale of Smiley's operation, but I didn't want him to think that I was a pale urban wimp from the air-conditioned North. If he wasn't wearing much protection, then I wasn't going to either. Never mind that these were his bees and he knew them well and had years of experience reading their mood. I tucked my shirt into my jeans, plopped a veil rakishly onto my head, disdained the gloves he offered me, and followed him through his hive inspection, showing him how tough and experienced I was. It was mid-April, just before the tupelo buds were expected to open, and Smiley was inspecting a row of hives that were gearing up for the nectar flow. The first few boxes went well. Smiley talked to his bees and told stories as I leaned in close to watch. I felt as though we were a couple of wise old beekeepers, quietly at work.

At the next box, perhaps the fifth in line, the volume and tone were immediately different. It seemed as if the very second that Smiley took the lid off, I felt a storm of icy burning pain on my knee and my left hand. My jeans had a tear a nickel wide at the knee, and when I looked through it there were two stingers throbbing on the little pink patch of exposed skin. With shaking hand and loud bravado, I pushed them out, and two from the back of my hand. "That was a queenless hive," Smiley explained calmly. "Bees can get aggressive when they don't have a queen." I nodded as I deliberated between crying and laughing from the painful, embarrassing surprise. Smiley, meanwhile, was wiping stingers off his arm as if they were drops of water and trying hard, I think, not to laugh at me.

By that evening my stung knee looked like my thigh. My hand was a useless hunk of blushed, furious flesh that seemed about double its normal size. All that long sleepless night, as I iced and elevated my giant, feverish limbs, I thought about my arrogance in assuming that all bees and hives were alike and that I knew all there was to know about basic beekeeping. The next day I cradled my left arm like a hot loaf of bread as I limped through the yards, keeping a respectful, wimpy, city-slicker distance from Smiley and his bees. Antihistamines, meat tenderizer, honey, wet tobacco, ice, and baking soda have all been suggested to alleviate the pain and swelling of a bee sting. I've had the opportunity to try them all, with minimal relief. I think the only things that work are patience and humor.

SIX

Food, Wine, and Fishing

*The principal things necessary for the life of men are: water,
fire and iron, salt, milk, and bread of flour and honey, and the
cluster of grape and oil and clothing.*

Ecclesiasticus, 39:31

*Thy lips, O my bride, drip as the honeycomb.
Honey and milk are under thy tongue.*

Song of Songs, 4:11

*Each time I cut a dripping square of wild honeycomb and eat it,
wax and all, I marvel at its perfection, which no processing
could possibly improve.*

Euell Gibbons, *Stalking the Wild Asparagus*

By July, the gallberry bushes have gone dry and slack in the summer heat, and other nearby nectar resources are similarly exhausted, so Smiley has moved all of his hives an hour north to take advantage of the later-blooming plants in Jackson County. The bees will snack on soybean, peanut, and cotton, and dabble in wildflowers, goldenrod, and fall asters. He will leave them feasting up there, with a honey super for each hive, until December, when the pink tentacled buds of the red maple trees in Wewa offer enough nectar to support his bees and he can bring them back down south.

Commuting to work now takes even more of Smiley's time, driving an hour each way to make sure his distant livestock is safe, healthy, and happy. During these visits (which he calls drive-bys), he checks for pests and other intruders and makes sure his bees have enough food. The rare colony that looks undernourished gets a ration of sugar syrup, fed into the hive like an IV, to prepare and nourish it through the cold, barren months of fall. Throughout July and August he visits each of his fourteen northern yards once a week, checking them with the fervor of an athletic coach making sure his players are resting and bulking up for the next season. "I have to know what's going on in those hives all the time," he says. "You can't do it by sitting in the house or going fishing."

When he is satisfied that his bees are content in their fall locations, he sits in the house or goes fishing. In Wewahitchka in August, it's too hot to do anything else. The average temperatures are over ninety degrees by day, accompanied by an intense inland humidity that makes it seem much hotter. Downtown Wewa is quiet and deserted, as though everyone has fled to the balmy coast or the air-conditioned malls of Panama City. Residents stay inside or sit deep in the shade of their porches watching the heat. Dogs kick up fine red dust before curling into sleep under the shelter of massive old trees. Everything and everyone seems to be waiting for a nonexistent breeze or the cool relief of September.

Inside the pink house, Smiley, Paula, George, and George's wife, Nancy, have been painting. They've applied shades of taupe and cocoa to the entry, living room, and kitchen. "I don't like painting much," says Smiley. "George doesn't care for it either, but he sure is good at it." Gearing up for a joke, he adapts the teasing drawl he uses for people he loves. "Why, George can take one drop of paint and paint a whole wall with it. Of course when he gets through, you can still see the darn Sheetrock." He laughs at his own joke. George, his hair highlighted with taupe, harrumphs and continues painting. Their efforts will be accented by the pinkish beige floor tiles Smiley and Paula have spent hours picking out in the chilly malls of Panama City. "I must have looked at a thousand and one pieces of tile," Smiley drawls. "Paula just had a ball picking out all the colors, but to me, if it doesn't have anything to do with bees or the bee business, I'm not that into it." He is into fishing, though. Bees and fishing. When the walls are painted and the days are still hot and languid, he hopes to take a day or two off on the Gulf or the river.

Ten minutes from the stultifying heat of town, the Apalachicola is a cool summer playground. The river and its tributaries churn down from Georgia, lazily passing through acres of tupelo swamp before emptying into the Gulf of Mexico thirty miles south of Wewa. Depending on sunlight or clouds, the silty, opalescent water is the color of split pea or lentil soup, but always as cool as vichyssoise. Branching off the river are myriad alleys of silvery tupelo trunks, a mirrored maze of watery green corridors that are shaded, cool, and noisy with wildlife. Elegant white egrets strut and pluck their way along the tangled banks. Black snakes and fat alligators sun themselves, camouflaged on semi-submerged logs. Copper-colored water snakes ribbon across the surface of the water, dissecting the diamond patterns raised by flitting waterbugs. Fish occasionally jump and plunk back into the water, rippling the dappled surface of reflected trees.

These endless secret corridors, rich with wildlife and the smell of fertile decay, were Smiley's childhood haunt (and the storied haven of bootleggers and criminals), and he does his best to haunt them still, taking his 13-foot fiberglass boat out whenever he gets the chance. "I just love it here," he says, admiring the tupelo buds and a sunbathing alligator. "When I want to get away, I just come here and sit and enjoy it. I should just get myself a houseboat and live out here." On the hottest days, many Wewahitchkans live on the river. Parking lots by the boat ramps are packed with trucks and empty trailers by nine in the morning. Leisure time is spent on the water, angling in boats or lounging at the elaborately terraced camps and houseboats that stud the banks. These rickety wooden shacks are little more than screened-in rooms perched on stilts or floats, where cool river breezes can be captured and lunch and dinner can be caught from the back porch. "We love our river, and we take care of it," says Smiley. "Most of us grew up on it. We learned to swim and fish and hunt on it. The river is the life of this place."

As a child, Smiley spent days and nights on the banks of the river gigging frogs, stalking squirrels, turtles, alligators, and deer, and catching fish. It was on the river that he was first introduced to beekeeping. "I remember walking through bee apiaries on the river to fish the banks just down from where I live now," he recalls. "I would be sitting and fishing and watch the bees at the water's edge collecting water, and I remember hearing the buzzing echo through the river swamp as the bees worked the tupelo trees. The only way to get to the river where I could fish was to walk right in front of

A tree-lined corridor of Apalachicola River swampland.

the beehives. All I knew about honeybees back then was they made honey and they would sting, so I just kept walking."

Although Smiley eventually slowed down and fell in love with the bees, he still adores fishing. "Before I became a busy beekeeper I was on the river almost every day," he says. Now he goes whenever he gets a break from his day job. He sets lines for catfish and casts a fly for bream and bass, tracking them the same way he does bees on a nectar flow, learning their habits and rhythms in order to maximize his yield. "Fishing and beekeeping are a lot alike," he says. "There's a lot of guesswork in both, and no guarantees. That must be why I like 'em so much."

He combines his two passions in the kitchen, drizzling the proceeds from a beeyard onto his piscine catch. After a day of fishing on the gulf, he lights the grill and fillets his catch with the same smooth expertise with which he smokes and inspects a hive, building a pile of gleaming fillets. Across one side he smears honey, adds a dash of salt and pepper, and flashes it onto the grill, which is housed in half a honey barrel in the back yard. He leaves it there for about a minute before taking it up to honey-baste the second side. Another minute later, the fish is done, lustrously firm and browned with just a taste of tupelo sweetness. His river catch he fries and drizzles with honey. Frisbee-sized frogs that George catches on the river at night might complete the feast. Smiley efficiently guts, beheads, and breads them before dropping the whole bodies into hot vegetable oil. To a store-bought hush puppy mix he favors, he adds sautéed onions and honey and dollops the puppies one at a time into the bubbling oil. Each will be plentifully drizzled in honey. Fresh local corn and a green salad (with honey in the dressing) complement the meal, accompanied by cold light beer and honey-sweetened iced tea. "I use honey in all my cooking," he says. "We don't own a sugar bowl."

In culinary tastes and lack of a sugar bowl, Smiley has a lot in common with the cooks of antiquity, who did not have sugar and also put honey in most everything they prepared. As the only pure sweetener known for centuries, it was a luxury drizzled onto and into just about every meal of the rich who could afford it. Some sugars could be wrung from fruits such as raisins, figs, and dates, but these were a diluted indulgence. Potent sweets and pungent spices were an extravagance enjoyed by the upper classes, while the less fortunate made do with bland, honeyless porridges, bread, cheese, and eggs, and occasional fruits and vegetables.

Apicius, the Roman epicure who lived in the first century B.C., compiled the earliest known and surviving cookbook, *De re Coqui-*

naria, or *On Cookery*, and over half of his hundreds of recipes employ the nectar of the gods. As one of the richest men in Rome, he could afford to indulge his sweet tooth. He spent his entire fortune on exotic food and lavish entertainment, and when his money ran out it is said that he committed suicide rather than be forced to economize. Luckily his recipes survived, offering a taste of the classical pantry and demonstrating the luxurious prominence of honey.

Apicius and his guests did not think of sweets and savories as opposites the way many cooks do today, but rather, as Smiley does, in combination. For cooked fish, he suggested the following salty/ sweet sauce: "Pound pepper with honey, lovage, thyme, oregano, rue, Jericho date, put it into a small vessel and add chopped hard boiled eggs, wine, vinegar, and the best oil." For ham, he instructs, "Boil the ham with plenty of dried figs and three bay leaves. Remove the skin and make criss-cross incisions, which you fill with honey. Next make a paste of flour and oil and cover the ham with this. When the paste is baked take out of the oven and serve as it is." The dormouse, a small rodent, was another sweet-and-salty Roman favorite when rolled in honey, smothered in sesame seeds, salt, and pepper, then roasted, although happily it hasn't maintained its popularity as well as the honey-baked ham. Desserts also displayed the Roman flair for sweetness and exotic spice. Apicius was fond of dates stuffed with nuts, pine kernels, and ground pepper, rolled in salt, and then fried in honey.

Honey was the first sweetener; it was also one of the first preservatives. At the beginning of *On Cookery*, Apicius advises on "How Unsalted Meat May Be Kept Fresh." Simply "Cover whatever fresh meat you wish to preserve with a layer of honey." The next suggestion is "To preserve skin of pork or beef and boiled trotters, add salt and honey to mustard prepared with vinegar. Completely cover the meat and use when you wish. You'll be amazed."

A hundred years later, Columella gave similar advice for preserving quinces:

> *Arrange them lightly and loosely, so they might not be bruised, in a new flagon with a very wide mouth.... They do not deteriorate any further once they have the liquid described above [honey] added to them: for such is the nature of honey that it checks any corruption and does not allow it to spread.*

Honey's power to preserve and to prevent decay comes from its sweetness. The solids in honey are 95 percent sugar, which kills most bacteria cells by osmosis, draining the fluid from them. Additionally, the bees add an enzyme to nectar during its transformation into honey that creates a small amount of hydrogen peroxide, making it an excellent preservative. With an average pH of around 3.9, honey has the acidity of a mild vinegar, another bonus in decay prevention. Honey is hygroscopic; it sucks moisture from the atmosphere and retains it like a sponge, which also contributes to its moist longevity and amazing preservative abilities. Honey farmers keep their product (with its average of 17 percent water), carefully sealed or in dry rooms, because in high humidity the thirsty honey will absorb an excess of moisture, and more than 19 percent will cause it to ferment. Beneath the surface, its density and saturation allow only a slow diffusion of moisture or dryness into the depths, so a sealed pot of honey will remain unchanged for months, even years, just two inches below the surface. Ancient meats submerged in it remained succulent and free from spoilage, and Romans, Indians, and Chinese kept their choice cuts sealed in amber pools of honey from one year to the next. Cakes infused with hygroscopic honey were lush and long-lasting. Suspended fruits and nuts retained their freshness and flavor for years.

Honey cannot stay moist or preserve matter indefinitely, though it has acquired that reputation. In *The Sacred Bee*, Hilda Ransome relays a frequently told tale of grave robbers discovering pots of succulent liquid honey in the ancient tombs of Egypt. "They came upon a sealed jar, which they opened and found it contained honey, which they began to eat. One of the party noticed that a man who was dipping his hand into the honey jar had hair on his fingers, and they discovered in the jar the body of a small child in a good state of preservation." While honey might have preserved the little bones for decades, it is unlikely that it could have done the job for centuries. But in the larders of history, it did the impressive trick of preserving foods from one year to the next.

Where kitchens were not advanced or available, honeycomb was a ready-to-eat meal, plucked from the nest and nutritious. With calories, vitamins, and minerals in every mouthful, it was a valuable high-energy treasure. Brood worms and stored pollen in the comb added precious protein. Many tribes in Africa still rely on these prehistoric MREs of honeycomb. The Mbuti of eastern Zaire continue to survive almost entirely off their glistening liquid catch. An observer of the tribe in 1981 noted that during one twelve-day period in the honey camp, 500 pounds of comb were gathered from a total of fifty hives. The twenty-three people in the group each consumed almost two pounds of comb per day, which was around 70 percent of their total diet. In his book *The Forest People*,

Colin Turnbull describes a typical Mbuti meal, unchanged for centuries:

> *Almost every hour someone returned from a secret forage, with leaf bundles tied to his belt, dripping the sticky liquid down his legs. Sometimes it was too liquefied to be eaten, and then the whole bundle was simply dropped into a bowl of clear forest water, making a sweet-tasting drink. But far more popular was the whole comb which could be eaten grubs, larvae, bees and all. If it was very hard, it was first softened over a fire, and this made the grubs squirm more actively, so that the honey worked its own way down your throat. It was, however, the best tasting honey of all.*

A continent away from the Mbuti, in the lands of the Old Testament, honeycomb was also a hallowed meal. Abraham and his descendants walked through the forbidding Syrian desert to reach Canaan, "a land flowing with milk and honey, the most glorious of all lands." When Jesus emerged from the tomb after his crucifixion, his apostles "gave him a piece of broiled fish, and of honeycomb. And he took it, and did eat before them." In India, the Buddha likewise broke his fast in the forest of Kosamba with a meal of honeycomb brought to him by a monkey. When the monkey saw that the vegetarian did not eat, he retrieved it, picked out all the brood worms, and returned it for the Buddha to enjoy.

Ancient civilizations were dependent on honey as food, sweetener, and preservative. They were also in awe of it: the sweetest thing that earthly tongues had ever touched, it was the taste of heaven. Although many of the exact rituals are obscure, it is certain that honey was a sacred meal or offering at almost all important spiritual and community events. Altars in Egypt, Rome, and Greece were slick and sticky with honey offerings, and rivers flowed with their fragrance. The pharoah Rameses III poured nearly fifteen tons into the Nile to thank and appease the river gods. Paintings in the tomb of Rekhmire depict life and death along the Nile, accompanied by offerings of honey. Honey was ritually "fed" to the Egyptian dead in a ceremony known as the "Opening of the Mouth," which enabled the departed to drink and eat in the afterlife. Because of its longevity and sanctity, it was the ultimate road food, stowed in abundance in the tombs for sustenance during the long journey to the next world. Pharaohs were entombed with their prized possessions, servants, supplies, and a loaded pantry that included honey in all its variations.

Honey offerings made for smoother transitions to the afterworld, and for sweeter earthly sacrifice. Sacred animals in Egypt, Greece, and Rome were fattened on the food of the gods, then stuffed with it for roasting and ceremony. Herodotus described a sacrificial feast of a calf when he visited Egypt in the fifth century B.C.

> *They stuff the body with hallowed bread, honey, raisins, figs, frankincense, myrrhe and other precious odors. These things accomplished, they offer him up to sacrifice, pouring into him much wine and oil. Then they beat themselves to appease the gods while the beast is burning on the altar, building up an appetite for the feast to follow.*

Births, weddings, feasts, funerals, even exorcisms were all solemnified and sweetened with honey. When Isaiah prophesies the birth of the savior in the Bible, he says, "Behold, a virgin shall conceive, and bear a son. Butter and honey shall he eat, that he may know to refuse evil, and choose the good." In India and Africa, newborn babies were similarly fed honey as their first meal to help ward off evil, choose the good, and ensure a sweet and healthy life. For the same reason, newlyweds in India fed each other honey for the first month after marriage and called the delicious period a honeymoon. Some say this term comes from the Middle Ages in England, when newlyweds were gifted a month's supply of honey wine, or mead, to ensure that married life began as merrily as possible. Wherever the word came from, it sprang from the universal notion that honey brought happiness and blessings to those who ritualized it. From Delhi to Jerusalem, temples, churches, and houses were erected on sites baptized with the sacred liquid. In Judaism, the New Year is still celebrated with apples dipped in honey, a wish for a sweet and blessed year to come.

Baptizing newborns with honey, on the other hand, is not so widespread or wise a practice anymore, as scientists have detected spores of *Clostridium botulinum*, which produce the toxin botulin, in some plant nectars. Common in nature and harmless to adults, the bacteria spores are gathered by bees as they forage and subsequently mixed into the honey in tiny and rare amounts. Before the age of twelve months, infants can't defend against botulin, so modern doctors and mothers think it best to hold off on honey rituals until the child's first birthday party.

In *The Sacred Bee*, Hilda Ransome describes a Babylonian ritual for the exorcism of a king:

> *The exorcizing priest anointed himself with a salve of honey and curdled milk. Seven altars were prepared in the palace yard on which were placed different kinds of bread, dates, meal, honey oil,*

butter and milk. After reciting a special formula the priest sprin-
kled a mixture of honey and butter to the quarters of the four winds.
He then went to an open field near the place where the seven ves-
sels were placed, and the king was washed with water. Sunrise was
then waited for and libations, in which honey played a part, made
to the gods.

Libations at these ancient ceremonies were often pure liquid honey, diluted honey, or the fermented version known as mead. Before wine and beer, before even an alphabet, mead was the intoxicant of gods and men alike. In his *Symposium* of 360 B.C., Plato describes the primacy, power, and potency of mead in the time before written words or wine:

When Aphrodite was born, the gods made a great feast, and among
the company was Resource the son of Cunning. And when they
had banqueted there came Poverty abegging, as well she might in
an hour of good cheer, and hung about the door. Now Resource,
grown tipsy with nectar—for wine as yet there was none—went into
the garden of Zeus, and there, overcome with heaviness, slept. Then
Poverty, being of herself so resourceless, devised the scheme of
having a child by Resource, and lying down by his side she con-
ceived Love.

Whether you were the goddess Aphrodite or a mere mortal, the powerful alcoholic ambrosia was easy to produce. With its abundance of sugar, natural yeasts, and a bit of added or absorbed water, honey acted as a convenient heaven- or home-brew mead kit. Colin Turnbull describes a chunk of comb in east Zaire that "was full of grubs too, but they were all dead, and the honey had fermented.... It tasted like a bitter liqueur: if eaten in any quantity it could be highly intoxicating." With or without grubs, the world's first cocktail was universally made from a simple combination of honey and water left to ferment for a few weeks or months. The result was a potent quencher of thirst and producer of hangovers; it also enabled glorious communion with the deities, because it was believed that a state of inebriation brought one closer to the gods (who shared a thirst for the magical brew). The Viking deity Thor is said to have consumed three tons of the "shining drink" in one sitting in Valhalla. When the consultation or favors of such divinities were sought, they were enticed and honored with an offering of a fermented honey drink. In the *Odyssey*, a feast is closed with thanks for and praises of mead: "Pour ye the drink offering, and send me safe upon my way.... And Pontonous mixed the honey-hearted wine, and served it to all in turn. And they poured forth before the blessed gods that keep wide heaven." Seeking the blessings and guidance of the gods, the Greek hero Jason shared a drink with them, splashing honey wine onto the waters beside his ship before setting out in search of the Golden Fleece.

Ancient Egyptian priests and kings drank and dispensed mead, as did the residents of Mount Olympus. In the sacred Hindu texts of the Vedas, the Gods of Light, called Asvins, drove a heavenly chariot full of honey and mead from whence they bestowed blessings on the mortals below, by means of a whip made of mead and honey. The song of the Asvins includes the following lines: "From heaven and earth, from the sea, from the air, from the wing, the honey-lash hath verily sprung. When the mead-lash comes, bestowing gifts, there's life's breath and there, immortality."

Grape cultivation began in Mesopotamia around 6000 B.C., and by 3000 B.C., it had spread to Phoenicia and Egypt. A thousand years later the Greeks were growing grapes and making wine, but they still drank a lot of honey. Oenophiles mixed honey and herbs into their growing wine selection, and by the first century B.C., an impressive variety of beverages was available in the kitchens, dining rooms, and banquet halls of the classical world. Palladius, who wrote a fourteen-volume treatise on gardening and agriculture in the fourth century A.D., described four different kinds of drinks made with honey. *Hydromel* was honey and plain water, *rhodomel* combined it with rose petals, *omphacomel* with fruit juice, and *oenomel* required honey, water, and grape musts. *Metheglin* was mead compounded with herbs. Along with honeyed wines these drinks were sipped for enjoyment and thirst as well as religious and medicinal purposes. Different concoctions were used to suit the pleasure, prayer, or ailment. Columella describes "a honey-water [that] can without danger be given at their meals to sufferers from fever. It is called melomeli."

Grape wine eventually replaced its honey-based ancestors in regions where vines grew easily, but to the north of the grape belt, mead was gulped, sipped, and celebrated well into the nineteenth century. Variations of honey wine were the drinks of gods, royalty, and commoners alike, and all consumed copiously. Much of the action of eighth-century *Beowulf* takes place in the Danish mead hall of Heorot, where the hero, "blithe with mead," feasts, drinks, boasts, and brawls. The hostess, "Helmings' Lady, to younger and older everywhere carried the cup, till came the moment when the ring-graced queen, the royal hearted, to Beowulf bore the beaker of mead."

At around the same time as Beowulf's epic in Denmark, it is said that a fire in Meissen, Germany, was extinguished with mead owing to a shortage of water and a typical abundance of honey drink. Two centuries later, Olga of Kiev arranged a conciliatory funeral feast with the enemies of her slain husband. She ordered quantities of mead enough to intoxicate and stupefy her 5,000 guests, and then, no longer conciliatory, she had them all killed. In 1460, the small town of Eger in Bohemia had thirteen mead breweries, which together produced almost four hundred barrels a year. Shakespeare drank mead, and his queen, Elizabeth I, had her very own royal recipe for it. In his diary, Samuel Pepys describes dining with the king in 1666, where "I drinking no wine, had metheglin for the King's owne drinking, which did please me mightily."

Pepys's King Charles II probably employed a royal mead maker to supply his table. He might also have extracted it as a liquid tax from some of his beekeeping subjects. Monasteries across northern Europe demanded tributes of mead as well, but it was often insufficient to quench the thirst of the church. Monks augmented their taxation supply by producing or purchasing huge amounts of honey for the in-house brewery. In one tenth-century monastery, two quarts of mead, which has the alcoholic kick of a strong table wine, were allotted to each monk at every dinner.

Common folk brewed mead at home. Typically, it was made from the leftover honeycomb after most of the liquid had been squeezed from it. The account books of Henry Best in Yorkshire explain how recently harvested honeycomb was transformed into the household beverage in 1641. "Into one tub the honey is wrung, and into the other, which should contain three gallons of pure water, goes all the dross of the hive, which is then boiled to make the three gallons of good mead which should be obtained from each hive."

Household guides from Best's era illuminate everything from plucking chickens to making cow brain casserole to methods of harvesting and using honey. They usually include several mead recipes ranging from a bitter, herb-laced medicinal draft to a sweet predinner cordial. *The Complete Country Housewife*, published in 1780 in London, instructs:

> MEAD, TO MAKE OF IT AN EXCELLENT QUALITY.
> *To six gallons of water add the whites of three eggs. When they are properly mixed, put to them 8 pounds of the best honey: when they have boiled an hour, put to them a little cinnamon, mace and cloves: let it stand till it cools; and then put to it half a pint of good yeast; when it has worked three days, let it be drawn into another vessel, and stopped close up for a month, when it may be bottled off, and it will be fit for use almost immediately.*

In northern Europe and the colonies of North America (where grapes were neither plentiful nor imported), honey was used to brew mead and to "renew," sweeten, and flavor ales, much as it had freshened and flavored the grape wines of ancient Greece and Rome. The *Dictionarium Domesticum* of 1736 suggests, "to renew beer that is flat or dead: Take four gallons out of a hogshead, and boil it with five pounds of honey—scum it, let it be cooled, and put it to rest, stop it up close, and it will make it pleasant quick and strong." The *Booke of Cookery*, a manuscript that was owned and used by Martha Washington for fifty years, offers similar beverage remedies for American colonists.

In South America and Africa, where bees have usually been more prevalent than grapes, honey-based fermented beverages were common for daily pleasure and ritual use. The Maya brewed a variant of mead called *balche*, using honey, water, and the pounded bark of a tree. The highly intoxicating (some say hallucinogenic) results were used to communicate with the gods in a variety of ceremonies and celebrations. According to Eva Crane, honey beer in tribal Africa "was essential for fulfilling many social obligations. It was carried to chiefs as tribute, and used to reward labor, an abundance of it was the glory of a chief's court or of a commoner's hospitality. Without it, tribal councils could not be held, and marriage or initiation ceremonies did not take place." In Ethiopia, the eligibility of a prospective husband was determined by how much honey he could offer as a bride price. His liquid payment was essential in making the *tej*, or honey beer, that tribes consumed in quantity on most occasions.

While it is still part of many rural traditions in Africa and South America, the making and drinking of mead and honey beer in Europe and North America tapered off to a mere trickle by the end of the nineteenth century. The increasing availability and affordability of sugar and a general shift toward cities, towns, and rum diverted people from the country necessity and tradition of producing honey and its ancient brewed by-product. The first sacred libation of men and gods became, by the twentieth century, a quaint memory of the rural past.

Honey was an offering in both its pure and fermented form. It was also rendered into a solid chewy version, in which it was combined with flour and spices and kneaded into fragrant dough that was cut, pressed, and molded to produce a variety of symbolic shapes. Dense, spicy devotional honey cookies were used in the same ways and ceremonies as pure liquid honey. By the fifth century B.C. the Greeks had over eighty shapes and varieties of bread and cakes made with honey and used in specific rituals. The *mulloi*, made of wheat flour, honey, and sesame seeds, was a detailed model of the female genitalia. Paired with an elaborate honeycake penis, both were offered to Demeter, goddess of agriculture and fertility at Thesmophoria, the ancient Greek Thanksgiving. There was a cake for every occasion in Greece—sacred, sexual, and mundane. In *Modern Greek Folklore and Ancient Greek Religion*, written in 1910, John Cuthbert Lawson describes the centuries-old custom of honeycake offerings. In Sparta, he writes, the peasant women

> will undertake even a long and wearisome journey to lay a honey-cake in a certain cave on one of the eastern spurs of Altgetus... the peasant-women consult the fates: with the girls matrimony, with married women maternity is the perpetually recurring theme. Everywhere also honey in some form is an essential part of the offering by which the fates' favor is to be won.

Honeycakes in all their flavors, shapes, and applications spread from Greece into Rome. Apicius recorded a recipe for a common honeycake known as *Liba Sigillariata*. To make the "consecrated cake," he kneaded together a pound of honey and a pound of flour, then added a spoonful of lemon juice and some pearl-ash (a rising agent) dissolved in water. After more kneading, the dough was left to rest for a day, then pressed into clay molds, shaken out, and baked. By the beginning of the first millennium, over three hundred bak-eries in Rome were churning out a variety of daily and occasional breads made with honey. Praising the skills and variety of the Roman confectioners, the epigrammatist Martial proclaimed, "This hand will construct for you a thousand sweet shapes of handicraft: the thrifty bee works only for him." Roman warriors were known to wear sweet honey dough amulets for strength and victory in battle, and cakes shaped as emperors and gods of victory were placed on altars or gobbled at feasts upon their return. At weddings, heart-shaped offerings were a popular treat. Recently, an archaeological dig in Rome unearthed over four hundred earthenware cake molds dating from around A.D. 200.

Ginger was a dominant flavoring for the spicy morsels, and through the centuries honeycake and gingerbread variations often came to mean the same thing. By the fifteenth century, most of medieval Europe had a highly developed gingerbread industry. In France, honeycakes were known as *pain d'épice*, or "spice bread"; in Germany, *lebkuchen* or *honigbrot*; in Russia, *prianiki*; and in Italy, *panforte*. In each of these versions, the devotional biscuits were also medicinal. Their abundance of herbs and spices, and the powerful infusion of honey made the cookies popular digestives and health aids. *The Dictionary of Science, Arts and Trades* of 1763 prescribed spice bread for upset stomach, and for abscesses on the mouth and gums: "Cut a slice of pain d'épice of a desirable weight and thickness: soak it in warm milk and apply it to the inflamed tumors."

The complexity and craft of these few square inches of cookie were a source of local pride, and many villages and bakeries were famous for their elaborate honeycakes. It was customary to mold detailed biblical scenes, folk and mythological characters, as well as the celebrities of the day in dense and savory honeycake or gingerbread. The spicy tokens could be eaten, kept on display, or given away as a delicious, edible commemorative. When Holy Roman Emperor

Frederick III visited Nuremburg in 1487, he had honeycakes molded in his image and distributed like royal trading cards amongst the people. Likewise, Queen Elizabeth I commissioned likenesses of herself and all of her courtiers in elaborately decorated gingerbread.

A honeycake mold commemorating the coronation of Matthias II of Hungary in 1608.

Medieval cakes, like the ancient ones, were offered and eaten on any special occasion. A bride might receive one in the shape of a baby as a token of good wishes. An unmarried woman could eat a man-shaped cake for luck in securing a husband. Many honeycake offerings were clustered around the Christian calendar, so angels, saints, and icons were eaten on feast days or hung from fir trees at yuletide. Sheets of fancifully decorated gingerbread were used to make miniature houses, which represented the sweet home and hearth of the coming year. Gingerbread men and houses come to Christmas mantels from medieval Europe by way of the honeyed altars and apiaries of ancient Greece and Rome.

In the American colonies, as in Europe, honeycakes commemorated and celebrated a variety of events. On election days, cookies were molded to resemble politicians or stamped with patriotic themes. "Training Day" or "Muster Day" cookies were handed out to militia and family members during the army's training exercises. When the circus was in town, the event was celebrated with sweet cookies and crackers shaped like lions, tigers, and bears. The *Booke of Cookery,* owned by Martha Washington, indicates how the cake would have been made in Virginia in about 1750. When shaped and dried, this dough became a gift, a digestive, a dessert, or an army of gingerbread men.

TO MAKE GINGER BREAD.

Take a gallon of ye purest honey and set it on ye fire till it boyle, then take it off and put into it allmoste halfe a pinte of good white wine vinegar, and it will make the scum rise yet you may take it off very clean. And when it is scumed put into it a quart of strong ale, and set it on the fire againe. Then put in halfe a pound of ginger, halfe a pound or more of good licorish, halfe a pound of anny seeds, 6 ounces red sanders. Let all these be finely beat and searced and mingle them well together, and let ye spice boyle in it. Then put in a peck of grated bread little by little, and worke it well in. And then mould it in searced cinnamon, of which you must allow halfe a pound to this proportion. When you have worked it well together, then print it in molds or make it into what fashion you please.

When sugar and its by-product, molasses, became cheaply available, they began to replace honey in baking. Eggs and leavening were also added, and gradually the ancient moldable chewiness of honeycakes gave way to an airier, more breadlike version. By the end of the nineteenth century, the cakes had all but disappeared from bakery shelves and folk traditions in North America and Europe. In 1873 there were as many as 2,000 honeycake makers at work in Hungary. Forty years later, when sugar was ubiquitous and cheap, there were fewer than twenty.

The history of honey's reign and abdication in kitchen, cup, and cake is the story of the advent of sugar. When he invaded the Indus Valley in 510 B.C., the Persian emperor Darius noted the locals cultivating "a reed that gives honey without the aid of bees." The conqueror took this discovery home and cultivated and hid it there until Alexander the Great arrived to plunder it one hundred and fifty years later. Alexander took some "reeds" and exploited their sweetness throughout the Mediterranean, to great wonder and acclaim. Dioscorides, a first-century Greek physician, sampled the "Indian salt" and noted, "There is a kind of solid honey called saccharon, which is found in the reeds of India and Arabia and the fortunate. It resembles salt in consistency, and crunches in the mouth." Thriving trade routes took the sweet discovery to the Middle East, Africa, and the rest of Asia, and by the time Marco Polo arrived in China in the thirteenth century, he observed numerous factories milling cane into crunchy saccharin crystals.

In the eleventh century, returning Crusaders brought the new spice back to Europe from the Middle East. Rich cosmopolites in Venice, Paris, London, and Rome clamored for the foreign sweet, and by the thirteenth century sugar was an obsession for those who could find and afford it. Sweeteners, because of their initial rarity and expense and our insatiable cravings, have been a status symbol for most of history. Just as early man had boasted with hoards and feasts of honey, he now showed off with a stash of sugar. A chic connoisseur's item, sugar was purchased and traded in little loaves that wealthy people carried in ornate silver boxes or gave as prestigious gifts. Cookbooks, written for and by the rich, called for sugar in all their recipes, while honey was suggested only "in default" of the stylish crystals. Supplies were limited, production distant, and prices outlandish, which only increased the cachet. King Henry III of England had difficulty obtaining just a few pounds for a banquet in 1226. In London in 1319, sugar sold for the equivalent of fifty dollars a pound. In 1515 a loaf the size of a little finger bought a lavish dinner for two in Paris.

As cane mania reached a crescendo, it was discovered that beets had a fair amount of sugar in them. In 1747 a chemist devised a way to extract this sweetness and produce crystals that were indistinguishable in taste from their cane cousins, ending the monopoly of the Indian salt. Production of the alternative crystals took off, cane sugar prices tumbled, and by 1775 a pound of cane or beet sugar cost a third of what it had in 1575. By the middle of the nineteenth century, honey and sugar prices were about equal, and by the end of it, sugar was far cheaper. At last within reach of the common cook, sugar was enthusiastically sprinkled wherever honey had once been sparingly drizzled. Sugar was new and exotic. It was also potent, dry, lightweight, and portable, and thus far more desirable than its messy liquid predecessor. Bees were suddenly passé, though they had been producing sweetness for centuries.

Just as it was suffering from the onslaught of cheap sugar, an improvement in honey production proved an ironic setback in its consumption. Hruschka's invention of the honey slinger in 1865 meant that for the first time liquid honey could be obtained relatively free of wax and other impurities. It was a tough sell. The slinger produced a liquid that was so clean, light, and clear that customers simply didn't believe it was honey. In *The History of American Beekeeping*, Frank Chapman Pellet describes the efforts of the Dadant Company to sell extracted honey in Illinois in 1870:

The trade, long accustomed to the sight of honey in the comb,
refused to have anything to do with the liquid product and refused
to believe in its purity. Many believed that they were manufactur-
ing the product instead of getting it from the bees as they previously
had done. The worst slander was from a local man who thought
that he could not eat honey. He stated that honey made him sick,
but that he liked the stuff the Dadants made. He said that he did
not know what they made it of, but probably good sugar, and that it
was just right for him. In the face of such an attitude, it required
much tact and patience to convince the public and dispose of the crop.

To complicate matters, genuine adulterers added sugar syrups
to honey to cheapen it, further heightening the mistrust that it was
a "pure" product. Crystallization posed another problem for honey's
credibility and reputation. In the comb, which is where most people
had heretofore obtained their honey, the sugars and water are in a
stable, saturated state. Extraction can cause the liquid to dry a little
and expose it to tiny impurities, forcing sugar to unsaturate and crystallize on the surface. Heating it gently reincorporates the crystals,
but devotees of old-fashioned honey did not know this. Suspiciously,
they saw granulation as further proof that the "new" honey was
fraudulent. Food adulteration laws passed in 1906 got the syrup
people out of the honey business, but consumers were slow to regain
their confidence in honey, and by then sugar was the cheap, reliable
victor. By the beginning of the twentieth century, honey cost almost
ten times as much as sugar, and honey dishes had been permanently
traded in for sugar bowls. *The Boston Cooking School Cook Book*
(which later became *The Fanny Farmer Cookbook*), published two
thousand years after Apicius's first honey-laden tome, categorizes

honey as "syrup" in the index and calls for the suspicious and archaic ingredient in only a dozen recipes out of several hundred.

The supreme reign of the sugar bowl has gone unchallenged for the past hundred years. Statistics from the United States Department of Agriculture indicate that Americans consume 152 pounds of refined cane and beet sugar and high-fructose corn syrups every year, which is about 52 teaspoons or more than a cupful every day. Honey, the ancient and supreme food of gods and men, has been relegated to a quaint condiment, consumed in this country at an average annual rate of about two teaspoons per person per week. That's just over two cups, or as much as a large can of soup. Some people go through more Nyquil in a year. But there is hope for a honey renaissance.

Those two teaspoons of honey cost three or four times as much money as the same amount of sugar, but price is the only advantage to eating crystallized cane. Each teaspoon of sugar is pure refined crystallized sucrose, a complex sugar that offers fifteen calories, a generic bland sweetness, and absolutely nothing else. Any interesting flavors or plant nutrients have been removed in the refining process. Honey, on the other hand, emerges from the hive refinery full of character, flavor, and natural organic nutrients. Depending on the floral source, a teaspoonful of honey provides twenty calories as well as antioxidants, amino acids, vitamins, and minerals including thiamin, riboflavin, niacin, vitamin C, calcium, iron, zinc, potassium, magnesium, selenium, copper, and manganese. Raw honey also contains trace amounts of pollen and its protein benefits. You would have to eat an awful lot of raw honey (I calculate about 7.5 cups!) to achieve the recommended daily allowance of these benefits, but you could eat sugar all day long (and it seems many people do) and still get nothing but a sweet processed buzz.

Honey is 40 percent fructose, a simple sugar that most taste buds experience as sweeter than refined, complex sucrose. For this reason, smaller amounts of honey can replace sugar in most recipes, while adding a complement of vitamins and minerals and the blessings of nature and the gods. It is more interesting to cook with the tawny food of the bees than with white processed granules, and it is certainly a more satisfying and sensual experience to tilt the honey bear, with its million floral nuances and variations, to your lips than a bottle of Karo corn syrup. Every gulp of raw honey is a distinct, unique, unadulterated medley of plant flavor; a sweet, condensed garden in your mouth. Alas, it will never be the only sweetener again, but honey may regain some favor and status among those who understand its benefits, history, and natural charisma.

For Smiley, the honey bear will always rule. Before beginning a day of drive-bys, he might have some leftover hush puppies for breakfast, or a bowl of corn flakes, either of which will be deluged with honey. Sometimes he makes a smoothie, using yogurt, loads of honey, and whatever kind of fruit he has on hand. "If I've got a raw egg in the refrigerator, that's going in the smoothie too. I use whatever I've got. One time I had some broccoli and I put that in," he remembers and pauses. "I won't do that again. But I put honey on just about everything I can. I put it in my gumbo while it's cooking. I put it in my cocktail sauce for the shrimp. I even put it on pizza." Apicius would have been proud.

TIME

(Or, bees and honey will outlast us all.)

New York City in August is as steamy and listless as Wewa-hitchka. New Yorkers stay inside and crank their air conditioners up high, causing brownouts on a regular basis. On August 14, 2003, my clock stopped at just after four o'clock in the afternoon. After a week of sweltering heat, New York had gone from brownout to blackout. The first thing I noticed was the quiet, which was eerie enough to wake me from my writer's power nap. I sat in my apartment with the windows open and read and wrote and listened to the unusual sounds of silence in New York City.

When it got too dark in my apartment to read, I went outside to investigate. I learned from a man with a transistor radio that ten thousand square miles and fifty million people were without power in what was being described as the worst blackout in U.S. history. It was not obvious to me that it was any kind of crisis—it was more of a party. People were wandering festively around sticky, smelly streets in various stages of undress and drunkenness, celebrating the adventure and novelty of powerlessness. There were encampments of beach and office chairs on the sidewalk, and residents chilling wine in the remains of their ice supply. It was as close as New York City will probably ever get to Mardi Gras. I climbed the darkened stairs to the roof of my building and settled in to watch the sun set. As the light faded to a dull peach glow and the granite buildings turned lilac, I enjoyed the haunting, dramatic gift of sunset in a city that is always lit. As people walked and murmured down the dark blue river of the street below, the amber glow of candles lit up squares of window across the way, transforming the dark building into a playful jack-o'-lantern. I heard laughter and clanking silverware and smelled wafts of garlic as my neighbors cooked by candlelight. That morning I had awakened in one of the biggest, rowdiest cities in the world, and at the end of the day I found myself in a quaint, peaceful village. As dusk was settling in and helicopters hummed and darted overhead like giant dragonflies, I observed bees dancing about on the potted plants of the rooftop, nonchalantly working the nasturtiums and flowering vines that had grown there all summer. They did not seem to know that they were in the middle of a national emergency and seemed as at home and at peace in the blackout village as in the daily electrified roar of Manhattan. As if I were sitting next to my backyard hives, the bees reminded me that all was well. It was foraging business as usual. They were doing what they always do, placidly going about their nectar gathering as they have for the last fifty million years, unfazed and undeterred by a little human power grid failure.

When I descended to my dark apartment, I was hungry. Watching bees always seems to make me hungry, perhaps triggering some prehistoric sweet tooth. Like most New Yorkers addicted to restaurants and takeout, I don't often cook in my kitchen. As the outage and sweltering heat dragged into the night, I took off my clothes and took stock of my situation. Twenty-four dollars in cash on hand and a bottle each of gin, red wine, and scotch. An assortment of olives, capers, grated Parmesan, and Dijon mustard graced my refrigerator shelves, along with some heavy cream for my coffee. My cupboard offered a box of pasta, a can of tuna fish, some tea, hot sauces, herbs and seasonings, microwave popcorn, and oatmeal packets that were probably five years old. I also had about ten pounds of honey, my connoisseur's collection. At the library, I had just read about East African hunters who lived on nothing but honey on their long trips in the bush, and the Mbuti of Zaire, who subsist on it for weeks at a time. If the outage went on, I calculated that at the Mbuti rate of consumption, supplemented with red wine and scotch and a bathtub full of water, I could live happily for five hot days. I had the world's first survival food in abundance.

Return to power was reported to be imminent, so I knew I could squander my resources. Having just finished reading Apicius's cookbook, I decided to go Roman and temper my usually peppery tastes with the sweetness of honey. Wearing only my camping headlamp to light my work, I lit the gas stove with a match and soon put some linguine into boiling water. Finding some freeze-dried garlic slices in the depths of my cupboard, I soaked them in olive oil and added some shreds of Greek olive, lots of hot red pepper flakes, a dash of hot pepper sesame oil, and a thick coating of fresh ground pepper. Over this concoction I drizzled about a tablespoon of honey (I think it might have been Smiley's tupelo) to create a dense, peppery sauce. When the pasta was cooked, I tossed it in my creation, then splashed heavy cream and a bit of grated Parmesan on top. I don't know if it was the magic of the blackout or eating naked in candlelight, but it was one of the best pasta dishes I have ever eaten. The honey bound it all together in a rich, elegant complexity, mysterious, sweet, and full of culinary history.

When power returned the next day, I, like many New Yorkers, was relieved but also disappointed, in a way, that the adventure was over. But I was thrilled, as usual, to have learned another lesson from the bees. Humans and their crises will come and go, I realized, but the timeless, unflappable dedication of the bees on the roof and the historic sweetness of the honey in my meal have not changed their tune or flavor in millions of years.

SEVEN

Liquid Currency

The more bees you have, the more money you make.
Donald Smiley

Nothing but honey is sweeter than money.
Benjamin Franklin

Weather and bees permitting, with 700 hives Donald Smiley can harvest and sell about 180 barrels or 115,000 pounds of honey in a really good year. With bulk prices for tupelo averaging around a dollar-fifty per pound, one might guess that he was becoming swiftly and sweetly rich in a town where the median income is $26,000. He's doing fine, but he's not getting rich. Each of his hives grosses about $150 in sales, but labor and expenses per colony can be as high as $100 annually. With healthy hives and clement skies, he nets about twice the median income of Wewahitchka. "When I was just getting started, with a few hundred hives, I was barely breaking even every year," he says. "Now I'm doing all right." To do even better, he wants more bees. To support more bees, he'll need more yards and equipment and a bigger honey house. "Right now I got a poor man's honey house," he complains with a smile. "It's too difficult to get the bottling done at the same time as you're harvesting. I really need a separate place to bottle my honey. Then I'll be a big-time honey producer." He's started clearing the wooded lot behind the pink house to make way for the big-time facilities, although he's not sure when he'll have the time or money to build. When the grand new honey house is finished, he'll use the poor man's for bottling and storage.

To finance his expanding operation, he makes sales calls to grocery stores, produce markets, and tourist shops in the Wewahitchka area and travels as far as sixty miles north to Dothan, Alabama, distributing his wares. Business is best during July and August, when tourists from Alabama and Georgia drive through the hot, sticky panhandle on their way to the relief of the forgotten coast beaches. In the height of the season, Smiley fills his truck bed twice a month and plies his route, keeping folks well supplied with souvenir Florida tupelo honey.

Souvenirs are made in the honey house when extracting is done for the day. At his stainless-steel bottling tank, Smiley stands and dollops twelve ounces of honey into plastic bears like an expert soft-serve operator at the Dairy Queen. He fills hundreds before proceeding to one-pound bottles, then quart containers and gallon jugs. Paula sometimes helps package the comb honey, one of the more popular products. In the kitchen, she lays a large slab of honeycomb on a sheet of waxed paper and takes a knife to it as if cutting a tray of brownies. Each dripping waxy chunk is nestled into a glass jar, where it will be submerged in liquid honey. Paula smoothes

blue-and-white or gold-and-brown Smiley Apiary labels on the filled
bottles and loops a festive "Florida Honey" flag around every neck
before warehousing it in the office or the extra bedroom. "I gotta build
a bigger house to hold all this honey," says Smiley, surveying the
operation.

One morning before he starts his route, Smiley's phone rings
at about 6:30. It's Curtis the barbecue man, in need of a delivery. Curtis is the proprietor of and cook at the Sea Breeze B-B-Que, a red-
and-white trailer parked in an empty lot in downtown Wewa,
across the street from Eddie's Beauty Salon. On Thursdays, Fridays,
and Saturdays year-round, Curtis sells mountainous piles of pork
ribs dripping a secret-recipe honey sauce. This Friday morning, he's
worried that he won't have enough honey to marinate and sauce
his weekend ribs. Happily for Curtis and his clientele, Smiley adds
the Sea Breeze to the top of the day's itinerary.

Two hours later, he pulls up in front of the cheerful trailer. The
screened and blackened smokehouse behind it wafts the delicious
odor of sweet wood smoke and roasted pork throughout the intersection. Curtis emerges with a twenty-dollar bill and a wide smile.
Smiley lowers the passenger side window and greets his customer
with a friendly wave. From the backseat of the truck's cab, a rum-
pled and sleepy George pulls up a plastic gallon jug of honey and
hauls it over to Curtis in exchange for the bill. Everyone exchanges
brief good mornings, and the first sale of the day is done.

Still smiling, Curtis heads back into the trailer to make his
secret sauce. Smiley takes a sip of coffee from the ever-present mug
and heads north up Route 71. Today, his sales and marketing uniform is a bright purple-and-green-striped polo shirt, black leather
belt, and stiff new blue jeans. On his feet are shiny white running

shoes and on his head a crisp new baseball cap, blue with "Alaska" in white embroidery. "I've always wanted to go to Alaska," he says. "It's the final frontier." George is also wearing a baseball cap, pulled down over a mottled head of rusty golden hair with dark brown sideburns. He was in the middle of bleaching his cropped cut the night before when a family emergency came up and he had to abort the process. The oddly orange results are a source of amusement for both beekeepers throughout the day.

Teasing his nephew lovingly and mercilessly about his new look, Smiley drives through fields of dark green collards and cotton, where the last beige bolls of the season droop on faded stalks. The sky is a moody slate hanging low on the landscape and threatening rain. Thick airborne battalions of dragonflies escort the truck for moments at a time before darting away. Black-and-red lovebugs, joined in flying coitus, bump and smear against the windshield. In spite of the rich fertility he is driving through, Smiley complains about the weather. It has been rough and unpredictable this summer, with a cluster of hurricanes hovering off the coast. The abundance of precipitation and shortage of sun caused the nectar to come late and sparsely. Cold and rain also made the bees ornery and reclusive, so it was often difficult for him to get work done inside the hives. "I'd get all the way to the yard, get the smoker started, and then it'd start to rain, so I'd turn around and come home," he sighs. The upside to his lament is that most local producers share it, so the crop will be smaller and prices a bit higher.

The second stop of the day is The Fruit Store in Alford, Florida. There is no sign of an actual town, just a wall-less emporium perched alone on the roadside. Signs leading up to the store and on the roof advertise LIVE GATORS, BONSAI TREES, GRAPEFRUITS, and

FLORIDA T-SHIRTS. Inside, the air is heavy, moist, and still, smelling of the damp wood-chip floor and earthy boiled peanuts. Pickled eggs, relishes, grapefruits, Vidalia onions, shot glasses, and shell collections are displayed for sale under bright fluorescent lights, and wind chimes hang soundlessly from the ceiling. George goes to look at Larry, Curly, and Mo, the live gators, who are sixteen inches long and hiss at him fiercely from inside their chicken wire cage. Smiley asks a silent older man slumped in a porch swing if the boss is around. A grinning woman named Suzy appears, and Smiley says cheerfully, "Hey, boss. I thought you might need a little honey."

Suzy eyes a shelf laden with jars of local strawberry-rhubarb jam, sorghum molasses, and Smiley's honey and says, "We still got a pretty good bit, I'm afraid. Business is way off." They discuss the recent jump in gas prices, the faltering economy, and ways to increase tourism and sales. Suzy says the tourists are crazy for the forlorn-looking bonsai trees she has displayed in a corner, and for Smiley's comb honey. "Mostly older folks," Smiley says with a nod. "They remember comb from the old farm days. Younger ones don't know what the heck it is." He makes a mental note to start her with more of the comb next season. Suzy wants him to build a display for his honey on top of some old hive boxes. He makes a mental note of that too. If you ask Smiley what he does with all these notes, he points to his Alaska hat. "It's all right up here."

At the next stop, the salesclerk, an elderly man with a damp white T-shirt stretched over his ample belly, phones his boss as Smiley enters the store. "The honey man's here," he reports to the mouthpiece. "What do we need?" The boss orders two cases of twelve-ounce honey bears and two boxes of the one-pound jars, which George arranges on the fake grass–covered shelves next to some watermelon rind pickles. With a borrowed sticker gun, he taps the price onto each container, $3.50 and $4.50 respectively, while Smiley and the clerk swap business and fishing stories.

Later in the day, at the Piggly Wiggly supermarket in Chipley, Smiley strides into the store with his arms wide and announces to no one in particular that he has come with honey. "You guys still got most of that honey?" he asks a startled-looking clerk with a pink-and-red pig embroidered on his shirt breast. "I don't know what we got," the man says, looking confused as Smiley advances, passes him, and goes directly to aisle six, where the honey is. The shelf is in disarray, with bears fallen sickly on their sides, and gaping holes where rows of honey should be. "Looks like a hurricane got in here," mutters Smiley as he begins to neaten and organize the shelves. "If there's empty space on that shelf, I'm gonna fill it with my honey," he announces as he makes way for his product. He sends his orange-haired deputy to the truck to get a supply of bears, which they mark at $2.89 each. The rows of Piggly Wiggly brand bears are priced at $2.79, and some other local producers are selling for as little as $2.50. "I'm only making a little bit off this packaged honey," he complains. "It's tough to compete with people selling for less than they should, less than store brand. They shouldn't be doing that."

At the end of a typical summer day on the route, they've made about twelve stops, delivered over eight hundred pounds of honey, and pocketed close to two thousand dollars, enough to pay some bills and buy some floor tiles and wall paint for the house. The produce markets took the bulk of Smiley's load, while the supermarkets seemed a little slow. "Honey always seems to sell better in grocery stores in the wintertime, I wonder why that is?" muses Smiley as he stows the receipts in the console next to his coffee. George pulls his hat down over his brow and says, "More people makin' biscuits." Smiley nods slowly, impressed with his protégé.

On the way back to Wewa, they discuss hair, beekeeping, and business. Smiley is planning on expanding his operation to more than a thousand hives in the next year, and hiring George full-time to help him. "To be profitable, you need to have a lot of bees," says Smiley. "If you got more than you can manage, you're losing money. George here is going to help me manage." He's grooming his nephew for the production side of things, so he can devote more time to selling. "I could stand on the sidewalk and sell honey, it's that easy," boasts Smiley. "I just don't have the time."

Second to the beeyards, the sales route is the work he enjoys the most, interacting with customers and proudly, boisterously promoting his product. His enthusiasm on the route brings in almost fifteen thousand dollars a year, a tenth of his total sales. Web site business (at www.floridatupelohoney.com) is growing at about 20 percent annually and will soon account for another tenth of his income. Not quite as much fun, the bulk of his sales are made over the phone early or late in the day, when the work in the yards or on the route is done. He answers inquiries, follows leads, and calls his regular roster of bakeries, restaurants, supermarket chains, wholesalers, and distributors. After a quick report on how the honey is coming in, he can usually take or confirm an order. Customers vary from year to year, but call by call, barrel by barrel, and bear by bear, he sells out.

Tupelo sells easily, first, and for a premium price. People always want tupelo. If any gallberry or bakery-grade honey is not spoken for, he can send it to Dutch Gold, the packing company in Lancaster,

Pennsylvania. If its price is right, he arranges for one of its trucks to come to Wewhitchka for a pickup. On the designated day, Smiley has the barrels stacked in the yard ready to go when a white eighteen-wheeler with pictures of honey bears and the Dutch Gold logo emblazoned on its side rumbles down Bozeman Circle. A few minutes later, Smiley has a receipt for his departed produce, and the truck is on its way again, picking up honey from other local beekeepers like a giant bee gathering nectar for the hive. When the belly of the truck is full, it begins the thousand-mile journey north to Lancaster.

Twenty hours later, the truck arrives in a landscape that is perhaps the cultural and geographic opposite of Wewahitchka. Yellow signs with black icons of a horse and buggy dot the roadside, as do pretzel shops and antique bazaars. Rolling green cornfields cradle old, orderly farms with stone barns and stenciled facades. Nestled into the heart of this Pennsylvania Dutch countryside is the Dutch Gold headquarters, a trim, modern white stucco building tucked behind an old farmhouse. Sixty million pounds of honey are processed here every year before being portioned into bears, bottles, and bulk containers for sale to food manufacturers, restaurants, warehouse clubs, and grocery stores.

Like so many honey businesses, Dutch Gold began with a few thousand bees, a backyard, and a curious beekeeper. Ralph Gamber, the founder, had a heart attack in his thirties, for which doctors suggested physical activity and a hobby. Gamber combined the two and in 1946 purchased three hives of bees for $27 at a farm sale. Over the next several years the hobby grew into a passion and then a business as the three hives multiplied into two hundred. With the help of his family Gamber harvested, bottled, and sold honey whenever he wasn't working his full-time job at a food-packaging company. His children helped him make deliveries of pint and quart glass bottles until 1957, when their father designed a squeezable plastic bear to showcase and deliver the product. Winnie-the-Pooh was immensely popular then, in the world and at the Gamber home, and Ralph, a plastics enthusiast, thought it would be fun to incorporate the beloved icon into his sideline business. The bear was a huge hit and became the prototype for the legions lining pantry and supermarket shelves all over the world. By 1958 the container business had grown to such an extent that Gamber quit his job, abandoned his own hives, and became a full-time packager for other honey producers. His son Bill took over the booming business in the early nineties, and in 2002 passed the torch to his sister Nancy, who is now president of the company. Another sister, Marianne, runs Gamber Container, the enterprise that grew out of the original bear. Ralph died in 2001, but his widow Luella still comes in to headquarters every week to water the plants and visit with employees and friends.

Baby food jars, plastic honey bears, and dark prescription bottles arrive at Dutch Gold's offices daily, sent by beekeepers auditioning their product. Every sample is run through the in-house lab, which resembles a brightly lit doctor's office, stocked with gleaming equipment and jars of golden liquid samples. Flavor, moisture, and color are evaluated (drier and lighter are preferred), and the honey is screened for additives and chemicals. If the lab prognostics are good, tastes of honey are run across the discriminating and formidable tongue of Nancy Gamber, the principal buyer for the company. Passing this battery of tests, the producer earns a place on the roster of Dutch Gold suppliers and can send it honey by the truckload or the bucket.

When a truck arrives at the plant, barrels are unloaded and stacked in a warehouse with the acreage of a football field. Smiley's barreled gallons join a towering, multicolored maze of honey from all over the world. Place of origin, flavor, moisture content, color, and quality grade (which is a combination of the latter three) are recorded in wax crayon on the outside of each barrel in the labyrinth of drums. When the processing of an order begins, a forklift driver fetches the components for each custom blend from this carefully marked liquid inventory. Like a vintner blending grape varietals or a coffee producer mixing beans, Dutch Gold combines flavor, dryness, and color to give the client the exact blend, body, and honey bouquet desired. It also packages a line of premium "monoflorals"— such as orange blossom, alfalfa, or sage—that, like fine wines from a particular grape, vineyard, and year, have not been blended at all.

Barrels for each batch are transported to the hot room, which is like a giant oven with garage doors, situated in the heart of the warehouse. The ingredients are parked here and their temperature slowly increased to 140 degrees. Heating resaturates any crystallization that might have occurred and makes the contents more manageably liquid for the blending and filtering ahead. Although the drums are sealed, the temperature sweats scent out of them, and the air around the heating room is thick and warm with sweet whiffs of honey. On days when the particularly fragrant orange blossom is heated, the warehouse smells like a humid citrus grove.

From the fragrant hot room, barrels of heated honey are transported to an elevated stainless-steel blending tub, where they are unsealed and upended two at a time. Molten honey cascades into a square metal vat below, where lazily churning paddles gently fold and mix the gallons of amber liquid. Blended batches are then piped next door to the filtering room, where their temperature is increased to the 180 degrees required to push them through a final series of cleansing filters. Not enough to cook or caramelize, the temperature renders the honey smooth and thin as water, and as easy to work with. As it exits the filters, the honey is cooled before being routed to holding tanks, where it rests at the temperature of a luxurious bath. From here, the warm river of honey is piped to the bottling room, where a fleet of bears and bottles awaits its cargo. The Dutch Gold plant is the honey version of Willy Wonka's chocolate factory. A river of warm tawny sweetness flows through the cavernous place, plied by men and women in blue jumpsuits and white hair nets busily and cheerfully transforming golden tributaries into smiling plastic bears and shiny glass bottles.

About 40 percent of Dutch Gold honey is bottled or beared. Honey that is not packaged for retail sale is sold in bulk to food service companies and food manufacturers like Kraft and General Mills. For these customers, processed honey is shipped in "totes," 3,200-pound plastic pillows of honey caged in metal bars. For even larger orders, 45,000 pounds are loaded, still warm, into a gleaming tanker truck before journeying to become Honey Nut Cheerios or Nabisco Honey Maid Graham Crackers. Some heated tankers drive from Dutch Gold to a portion-packing facility that divides the twenty tons of liquid into millions of little gifts from the hive, the plastic cups and squeezable half-ounce envelopes that sweeten the fare in restaurants and coffee shops.

Every bulk container that leaves the Dutch Gold warehouse has a locked red plastic security seal, which guarantees that the contents have been tested for additives and impurities. In the year 2001, trace amounts of the harmful antibiotic chloramphenicol were found in imported Chinese honey. Beekeepers there had used it to combat an epidemic of hive disease but failed to keep it from contaminating the product, which unfortunately made its way into the global supply. Chinese honey was temporarily banned, but nervous buyers demanded further guarantees of purity and safety, and the seals came widely into use. Security concerns escalated with revelations that containers of honey were being used to traffic money, drugs, and arms throughout the Middle East, where honey vendors are ubiquitous. Thick, fragrant, honorable honey had been put to a dishonorable new use: smuggling. A U.S. government official was quoted in *The New York Times* as saying, "The smell and consistency of the honey makes it easy to hide weapons and drugs in the shipments. Inspectors don't want to inspect that product. It's too messy." When anthrax and tainted food supplies became a concern soon after September 11 of that year, security seals became regulation at Dutch Gold and other major dealers in the suddenly suspicious liquid. Even smaller suppliers like Smiley came under scrutiny. In 2003, complying with new bioterrorism regulations, he registered with the USDA, asserting the purity of his honey and his intentions. Backyard beekeepers making honey for their friends or the farm stand are thus far (and hopefully it will stay this way) unregulated and unregistered. They just give their solemn word that the product is innocent and delicious. And usually the honey speaks for itself.

From bears and bottles, totes and tankers, the National Honey Board estimates that Americans consume approximately 400 million pounds of honey each year, or about 1.3 pounds per person, twice as much as the 1,600 commercial beekeepers (defined as apiaries with over 300 hives) in this country can produce. Large distributors like Dutch Gold have to import almost a third of their annual supply from foreign producers, typically in winter, when American apiaries are dormant. Foreign honey, which is abundantly available and cheap, has turned into a year-round concern for beekeepers in the United States. Until the late 1990s, China and Argentina were the largest global exporters of honey (and two of Dutch Gold's biggest suppliers). They exported honey for around thirty cents a pound, a third of what it cost Americans to produce the same amount. These prices devastated American beekeepers, forcing many of them out of the business. "I was getting offers for forty cents a pound," recalls Smiley. "I had one hundred barrels of bakery honey sitting there, and I couldn't sell it because I couldn't compete with Chinese honey."

While Smiley was getting offers of forty cents a pound, the American Beekeeping Federation filed a formal complaint, claiming that the cheap and often inferior foreign product being dumped on the market was deluging and destroying the American honey trade. As a result, the U.S. Commerce Department began imposing taxes of almost 200 percent on honey from China and Argentina. Antidumping tariffs reduced imports from those countries to a quarter of what they had been, and prices for American honey more than doubled. Smiley and the market recovered, but not without future foreign worries. Tariffs are temporary, and cheap competitors are always waiting to fill the American honey deficit. The president of the American Honey Producers Association has said, "We can't produce enough honey for our own market, so we will always be an importing country. Beekeepers must understand that we are dealing with a world market and we are going to have to learn to live with it."

Tupelo helps Smiley live with the foreign worry. It is always in demand, fetches consistently high prices, and is not threatened by foreign competitors, who don't make it. It helps to live in Wewahitchka, tupelo Shangri-la. Thanks to the missionary with her seeds and the thief who stole and flung them, unique, plentiful acres of swampland and specialty nectar in his hometown are enough, if necessary, to take care of him for decades. "There are tupelo trees in some of the swamps so far in that no bee has ever touched them," he marvels. Most of the swamps and trees are in state parks, so Smiley's bees and future income are relatively well protected. "I can always sell tupelo," he says. "Tupelo's the thing that keeps me in business."

Smiley's list of business worries are: too much rain, too little rain, foreign honey, parasites, hive diseases, and pests. Smiley's list of personal worries are: too much rain, too little rain, foreign honey, parasites, hive diseases, and pests. Small hive beetles (*Aethinae tumidae*) are of African origin but made their way somehow to Florida, where they were discovered in 1998. Since then, they've proliferated like weeds and are now found desecrating hives throughout the American South. They are about the size of a ladybug but black, ugly, hairier, and much faster. Established in a hive, the beetles consider the colony's resources their own, consuming honey, wax, pollen, and larvae at a voracious, destructive rate. They are much less hygienically fastidious than bees and defecate in the honey, causing it to spoil and seep from the combs. Beekeepers know they have a bad beetle infestation when their honey smells and drips from the frame like rotten fruit juice.

Foulbrood is another common hive affliction, the result of bacteria that invade the hive and reduce the brood to a brown, stringy, useless soup. Two different strains of foulbrood, American and European, were the biggest problem in commercial beekeeping until foreign honey and mites came along in the early 1990s. First discovered in Indonesia in 1904, varroa mites made their way to the United States by 1986 by stowing away in worldwide shipments of bees. These ticklike parasites crawl into the bee brood cells, where they feast on the larvae and mate with abandon. When the adult bee emerges, if it is even able to do so, it is bedraggled host to several crippling mites. A colony can withstand a small mite invasion, but large marauding numbers of varroa will seriously compromise its strength and productivity. Another treacherous mite, the tracheal, sets up parasitic shop in the trachea of a worker bee, where it lays eggs and cripples the host's ability to fly and feed. Beekeepers recognize a tracheal mite invasion when they see bees limping and crawling about the hive, unable to report to work. While foulbrood and beetles can be treated, mites have become increasingly resistant to chemotherapy; dozens of chemicals have been applied, with little long-term success. Researchers are experimenting with essential oils, screens, and mite-resistant bees with hope and some success, but 20,000 colonies in Florida, or 8 percent of the state's commercial colony count, are still lost each year to the ravages of mites. If an effective defense is not found soon, the outlook for Smiley's bees and his livelihood, tupelo and all, is bleak. "Mites are here to stay," says Smiley. "They're going to be a constant battle from here on out."

With all of its irritations, uncertainties, and rewards, the honey business as Donald Smiley knows it has existed only since the late nineteenth century, when beekeeping innovations in Europe and the United States produced previously unimaginable surpluses. Until these advances, honey was primarily a domestic, limited, hive-to-mouth endeavor. Bees were kept close to the household, like a good laying hen or an exceptional milking cow, and their emissions were consumed within a few yards of production. In and around ancient Rome, residents embedded hives in the clay and stone fences surrounding their villas, and in India, North Africa, and Asia the mud walls of the houses themselves were studded with hives. Hollow log hives in central Africa and China were suspended beneath the eaves of thatch-roofed homes. As late as the 1850s in Europe, new houses were built with niches in the outside walls to accommodate skeps. Bees were an integral, intimate part of the household, their produce used to sweeten food and drink and illuminate the rooms within.

Discussing the best domestic arrangement for bees, Columella wrote in the first century that "It is expedient for the apiary to be under the master's eye." In ancient Greece and Rome, the master was often a male slave known as the *melitore*. Beekeeping was man's work, and Columella advised that, to conserve their strength for the manly task, bee handlers abstain from sex for at least a day before going into the hive. Women were considered downright dangerous when it came to handling bees. Pliny the Elder warned that "If a menstruous woman do no more than touch a beehive, all the bees will be gone and never more come to it again."

Later, the menstrual prohibitions seem to have faded away, and the task of caring for bees fell to the females of the house. John Levett wrote *The Ordering of Bees* in 1634, proclaiming in the foreword that "The greatest use of this book will be for the unlearned and country people, especially good women, who commonly in this country take most care and regard of this kind of commodity."

Eighteenth- and nineteenth-century houses in England
were built with niches for beehives.

The author of *The Complete Country Housewife* declared the bee-keeping particulars of his book "worthy of the attention of women of all ranks residing in the country." While wild bees and honey hunting were manly pursuits, the tending of bees, chickens, cows, and the household herb garden were typically the dominion of women. Eva Crane writes, "In general, human gender roles in relation to bees and beekeeping conformed to gender roles in other activities."

Skeps were part of the farm woman's chore list from the first through the nineteenth centuries. The *Dictionarium Domesticum*, "A new and complete household dictionary" published in London in 1736, features illustrations of a woman going about her household tasks. She feeds chickens, churns butter, launders, cooks, bakes, and brews beer. In one tableau, the woman sits at her kitchen table and the open back door behind her reveals three busy backyard skeps, as important to her ménage as the butter or the beer. One hundred and fifty years later, bees and homemade honey were still the showpiece of an efficient housewife. A writer described a garden hive with glass collection jars "of such a size as to suit a family to breakfast, each of which may be daily introduced to the table fresh from the hive. A little honey on bread would save the use of butter on the occasion, and would be more wholesome; it is at the same time a luxury, that every family in possession of a garden, may command without expense."

If they could, most households supported enough bees to supply the family with honey for the table and the mead vat, and wax for candles. A cottager might keep seven or eight hives, a larger household employed as many as ten or twenty, while grand estates had

From *Dictionarium Domesticum*, a housewife's
compendium and guide.

Illustration of a medieval Italian household and garden.
Beehives are housed in domed niches in the wall directly in front
of the house and some are also shown in the adjacent garden.

apiaries with dozens of hives. For most of their domestic history, bee domiciles outnumbered those of humans. *The Domesday Book*, compiled in England in 1086, inventoried everything of value in the king's domain, including hives. There were 1,441 registered in rural East Anglia alone, at a time when there were fewer than ten people per square mile. William Lawson's *The Country Housewife's Garden*, published in England in 1618, offered beekeeping advice "with secrets very necessary for every housewife" and concluded, "if you have but forty stocks, shall yeeld you more commodity clearely than forty acres of good ground."

In the American colonies, the number of hives under the housewife's command was similarly abundant. In New York, Daniel Denton noted in 1670, "You shall scarce see a house, but the south side is begirt with hives of bees, which increase after an incredible manner." A hundred years later, a soldier in Pennsylvania observed that "every house has 7 or 8 hives of bees." An estate for sale in nearby New Jersey included a house, barn, "valuable breeding mare, sheep, swine, several flocks of bees, household furniture, and farming utensils."

Seven or eight hives (which, until the nineteenth century, were about half the size of today's boxed hives) would have provided adequate honey for household baking, cooking, brewing, and sweetening. If there happened to be surpluses, they could be sold or exchanged for a variety of valuable goods and services, from bulls and bribes to brides and wine. The account books of the Egyptian pharoah Seti I indicate that one hundred pots of honey could be traded for an ass or a bull in the thirteenth century B.C. In the first century B.C., Diodorus of Sicily recorded honey being traded for protection from occupying Roman forces who were "lords of the cities for a considerable period and exacted tribute of the inhabitants in the form of resin, wax, and honey."

A marriage contract in ancient Egypt contained a promise by the groom to deliver twelve jars of honey to his bride every year in exchange for her hand in marriage. In Africa, a single payment of twenty-five pots of honey was given to Masai fathers for matrimonial rights to their daughters. Strabo's *Geography* of the first century describes Ligurian villagers who traded their surplus honey for olive oil and wine. In Strabo's world, honey was often valuable enough to be traded equally with salt, that most precious currency.

Until the nineteenth century, in many parts of the world the rent could be paid in honey. Tax and rent laws were drafted throughout Europe specifying how much sticky tribute the landlord and sovereign could exact from each of his tenants. Eva Crane writes that in England in the eleventh century, a fine was imposed on a woman in Somerset, who "for six years has withheld her rent in honey and cash." In some cases villages combined their honey resources and paid dues and rent to the local landowners and rulers. In South America, Mayan and Aztec warriors expected several hundred jars in honey annuities from their conquered foes, liquid rent for their own land. To sweeten his banquets (and his wallet), a greedy seventeenth-century Hungarian commissar depleted the honey supplies of the entire tiny town of Nagykoros, demanding 2,400 gallons in rent and taxes from the inhabitants. If only the IRS or MasterCard worked that way today.

When and where forest beekeeping was practiced, itinerant beekeepers often paid for the privilege by handing over dripping portions of the proceeds to the church and landowners. In parts of northern Europe, beekeepers tithed a portion of honey to neighboring farmers in exchange for letting bees forage on their property. When taking that honey home or to market, they would frequently have to pay a sweet liquid toll at the bridge or road into town.

As providers of the coveted sweet, bees were themselves valuable commodities. Knowledgeable hunters could capture swarms in the wild and convert them to a steady profit stream. Aristophanes and other writers of antiquity mention a separate area in

One of many examples of currency graced with bees and hives.

the marketplace where hives were sold. References to exact prices are scarce, but in 918 the worth of a new skep was the equivalent of two hundred eggs or twelve arrows. In the American colonies, damage claims after the War of Independence frequently inventoried bee losses, and one reported thirteen destroyed hives valued at forty-five dollars. Bees were so highly esteemed, and their honey so highly valued, that both were featured on local currencies.

In 1853, two hives of bees were sold for two hundred dollars to settle the estate of an American man who had been killed in a steamship explosion. A late-nineteenth-century photograph shows a bee market in Denmark, in which the smiling proprietors lounge next to a row of tied-up straw skeps, which they could sell for about seventy-five dollars each. Today, a three-pound starter package of bees with a queen, obtained from a dealer, costs around fifty dollars, and two assembled deep hive bodies with frames, bottom board, cover, and a super are another hundred and fifty. Tools to handle the bees—smoker, gloves, hive tool, and veil—will run to fifty or sixty dollars. Bee suits are optional (and extremely warm), but the security and confidence they provide the beginning beekeeper are definitely worth the sweat and the eighty-dollar price tag.

Though honey was typically a wild windfall or a limited domestic production, entrepreneurs have always procured surpluses for profit. While women usually tended the household hives, enterprising men established large apiaries to extract as much of the difficult yet lucrative yield as possible. Before 1865, however, manipulating "commercial" apiaries, unwieldy collections of logs, boxes, and baskets, was an arduous, sting-filled business. No matter how large and efficient the apiary or how expert its proprietor, yields were limited by knowledge and equipment to a few small, sticky harvests wrested from each hive each year. Langstroth's revolutionary adaptation of bee space and Hruschka's invention of the extractor in 1865 transformed honey from a domestic trickle or speculative sideline business into a thriving industrial flow. Other improvements helped. In 1869, not long after Langstroth's revelation, the first transcontinental rail route was completed. With this and various other innovations in transport, massive numbers of bees and tons of their produce could be shipped quickly and cheaply to apiaries or to market. Additional advances in woodcutting machinery and plummeting lumber prices converted makeshift rural apiaries into neat rows of uniform wooden boxes that served as efficient modern honey factories.

Honey was now mass-produced via an assembly line of cheap, interchangeable, reusable parts, and surpluses and profits in beekeeping became abundant and reliable. By the end of the nineteenth century, machines and modernity surpassed centuries-old traditions and manual labors. As innovations multiplied, honey became a modern "manufactured" good that could be traded by the truckload and the ton instead of by the jar or bucket. Smiley's vocation is thousands of years old, but the commercial honey business is just a century young.

Some of the biggest surpluses in the new industry have been realized in California. The state did not even have bees until the middle of the nineteenth century, but the climate and geography presented a rich vein of opportunity. With the excitement of miners in a gold rush, entrepreneurs imported bees to California for exploitation and riches. John S. Harbison of San Diego began receiving shipments of insects from the East by rail and in a few years was the premier honey producer in California. By 1876, the superstar beekeeper was able to ship a magnificently impressive and altogether unheard-of trainload of honey from his apiaries to New York City. In *The History of American Beekeeping*, Frank Chapman Pellett writes, "Harbison's shipment of ten cars of honey to New York in 1876 created a sensation in the East. Beekeeping at that time was

A southern beekeeper shows off his modern apiary in 1915.

generally an insignificant side line and few beekeepers produced as much as a ton of honey in one year."

A book published the next year claimed that Harbison had netted $25,000 on his 200,000-pound transcontinental shipment of honey, a considerable sum in 1876 and an unthinkable one just twenty years previously. With fame, wealth, and 3,500 hives, Harbison was one of the most successful apiarists to stake his claim in the California honey rush. According to the *American Bee Journal*, others were tapping into the western lode. "The crop of 1885 was about 1,250,000 pounds. The foreign export from San Francisco during the year was approximately 8,800 cases. The shipments east by rail were 360,000 pounds from San Francisco, and 910,000 pounds from Los Angeles." Today, California still generates the most honey in America and, in 2003, produced thirty-two million pounds of honey worth $41 million.

In Florida, the race was a little slower. By 1872, when the California rush was in full swing, Florida was still a ferocious, roadless frontier where the bee business was conducted with considerable difficulty, mainly by boat. Pensacola (about an hour from Wewa) managed to support one of the first commercial apiaries in the area, although beekeepers had to move heavy boxes by skiff through swamps rife with alligators, snakes, and mosquitoes. It wasn't easy working this landscape compared to the rolling green hillsides of California. In 1901, one of the pioneers in the area, M. W. Shepherd, described his precarious experiences in an article entitled "Beekeeping in West Florida, a Letter from a Land Flowing with Malaria and Honey."

Twenty years later, a Canadian beekeeper named Morley Petit spent a year attempting his trade in Florida, after which he decided to hang up his veil and return home to the relative tameness and ease of Ontario.

> *Outside of Apalachicola districts this [the Lake Worth area] is said to be one of the best honey-producing districts of the state; but with all its uncertainties, strange pests, and the long active season, the surplus is probably no more than we get in a short sharp season and have done with it. This northern beekeeper is satisfied to raise his honey and make his money in good old Ontario, where he knows what to expect.*

At about this same time, a beekeeper reported losing an entire boatload of bees to an unexpected squall in Pensacola Bay. Florida beekeeping was for swashbucklers, not the faint of heart. Census reports from 1910 indicate that the entire state of Florida eked out about 340,000 pounds of honey, close to what the city of San Francisco had shipped east twenty-five years previously. From its humble, hazardous beginnings in Pensacola and Wewahitchka, Florida, beekeeping (with the advent of citrus crops in the South) has grown in the last ninety years into the third largest producer of honey in the United States. Threats from malaria, alligators, and sudden squalls have lessened, but it's still not the easiest place to produce honey, due to frosts, drought, beekeeper attrition, and pesticide and mite problems. Production in 2003 was down 27 percent from the year before, to fifteen million pounds worth $19 million.

Donald Smiley is part of a rich entrepreneurial tradition, but he's not as rich as earlier entrepreneurs. In the time of *Domesday* or in the California glory days, seven hundred hives would have made him wealthy. He could have paid the rent and George in honey. In

A man and a young boy in the honey business.

Egypt he could have boasted several bulls, and in Africa he might have been able to afford many wives. In the worrisome business of modern honey, he has one beloved wife and, weather permitting, does okay. "I'm probably not going to get rich doing this," he says, thinking of the threat of foreign honey prices, and of the bees that need tending, the honey house that needs cleaning, product that needs selling, and a sticky pile of frames that require repairs. "It's a lotta work and a lotta aggravation. But I love it."

EIGHT

Wax

He must be a dull man who can examine
the exquisite structure of a comb, so beautifully adapted
to its end, without enthusiastic admiration.

Charles Darwin

We have rather chosen to fill our hives with
honey and wax; thus furnishing mankind with two
of the noblest things, which are sweetness and light.

Jonathan Swift

We must no more ask whether the soul and body are one than
ask whether the wax and the figure impressed on it are one.

Aristotle

I n October, between drive-bys, sales calls, and moving into the
new house, Smiley harvests yet another crop, wax. After a season of extracting honey, he usually has about twelve green metal

barrels full of wax caps ready for rendering. During the harvests, the green barrels fill cap by cap, inch by inch, as supers cycle through the honey house. Beneath the chains of the uncapper, shorn wax caps clump together and slide like sludge down metal ramps before dropping into a corrugated tub two feet below. The pile builds as every frame yields clumpy handfuls of shredded wax slathered in honey. Newer wax, which might have been on the comb for all of a few days before meeting the uncapper, is light and buttery white. Wax that has been in the hive longer has darkened with age and use to a light caramel color, and the oldest wax is a dark chocolate brown. As hundreds of frames pass above, the mound builds to a decadent dune of what looks like butterscotch-coconut cake icing— a fibrous concoction of golden honey and earthy wax shavings dotted with dark brown dead bee raisins. In the heat of the day the contents of the tub ooze together like rich sweet mud.

As the heap of gooey icing grows, George grabs a garden shovel and mucks out the tub, emptying it into a long metal trough a few a feet away. Over the course of a few warm days, most of the remaining honey will collect at the bottom of the tub and drain off, leaving wax that looks like thick wet barn sawdust. Throughout the busy summer, it is shoveled into clean metal storage drums, then sealed and set aside until quieter, cooler days.

On one such day in mid-October, Smiley mounts his bright red Bobcat and approaches one of the loaded barrels. He forklifts and conveys it upright across the yard and lowers it onto the top of his rendering apparatus, a rounded steel frame that holds the barrel above a propane gas burner. Using a battered plastic juice pitcher, Smiley pours five to ten gallons of water into the drum, then lights the burner at the base. He keeps the flame fairly low—more of a delicate poach than a boil, because too much heat can damage and discolor the wax. As he engages in other tasks around the yard, he periodically monitors and stirs his redolent stew. In four or five hours, all the shavings have melted, and the whole soup has subsided to a foot or so below the rim of the barrel.

Using the plastic pitcher again, Smiley ladles warm liquid wax, the color and consistency of a light maple syrup, into gray fiberglass molding pans that look like food service bus buckets and hold forty pounds of wax. They are left outdoors overnight to cool and harden. The next morning, he slips them into the refrigerator in the shed for a few hours to set the wax firmly. When he taps them out of the fiberglass shells, he has heavy golden bricks of pure beeswax worth about fifty dollars each. Good wax sells for about $1.25 a pound, as much as his bulk gallberry honey, and is an important part of any commercial beekeeper's business.

After the wax has been ladled off, a foot of dark detritus remains in the barrel. The top layer of this gunk is called slum gum—six inches of black waxy sludge comprising bee parts, pollen, honey, brood, a bit of wax, and a lot of dirt. The traffic of hundreds of thousands of bees, each with six foraging feet, tracks a surprising amount of debris into the hive and across the comb. When cooled, the slum gum resembles a waxy peat moss or a crumbling old brownie. Smiley hoards chunks of gum in a barrel. At the end of a season or two, his collection will be large enough to render into another brick of wax and another fifty dollars.

When all of the shavings have been converted into thirty-five bricks, Smiley heaves them onto his truck and drives five hours east to the Dadant Company in High Springs, Florida. Through its catalogue and branch locations, Dadant sells every kind of supply for every level of beekeeper, from novice to industrial. From the larger honey producers, Dadant also purchases rendered wax to be molded into comb foundation for sale. Usually Smiley runs into several other honey farmers at Dadant, each with a shipment of wax in his truck, each with a list of supplies to be exchanged for that load. Using over 1,400 pounds of high-grade beeswax as currency, Smiley takes home parts for 400 new deep supers, 4,000 frames to go in them, and 4,000 sheets of foundation. His wax dividends pay for the refurbishment of some of his older hives and for the construction of new ones for the spring.

A colony spent days, weeks, even months building the wax comb caps that Smiley scrapes off and melts down in a few hours' time. Bees produce the comb in painstaking increments, using tiny flakes of wax as both bricks and mortar. At only twelve to eighteen days old, a worker bee charged with construction hasn't yet ventured from the hive, but she is familiar with its innermost workings. When nectar income is heavy and the queen is laying eggs with determined speed, the worker senses the need to produce wax, the building material required in spring and summer when the colony is expanding. If a bee's life takes place during the fall or winter, when the colony is quiescent and new comb construction has ceased, she knows not to worry about building supplies.

When wax production is on the seasonal agenda, a bee will first gorge herself on nectar from a storage cell, like a sumo wrestler putting on pounds for a meet. After the binge, she will rest quietly for up to twenty-four hours, allowing the sugary excess to metabolize into wax. Lorenzo Langstroth wrote, "Combs are made of wax, a natural secretion which is produced by bees as cattle produce fat by eating." A colleague of Langstroth's marveled: "We put our poultry up to fat in confinement, with partial light, to secure bodily inactivity, we keep warm and feed highly. Our bees, under Nature's teaching, put themselves up to yield wax under circumstances so parallel."

Although the circumstances of its creation are parallel, beeswax is not pure lard but a complex mixture of more than 300 chemical components, many of which are fatty acids and esters. Bees don't store this substance as potential fuel and flab the way cattle, poultry, and humans do but utilize it in a much more creative way: to build homes, store food, and protect their young.

As the lethargy of her nectar binge is wearing off, the fatted construction bee moves into position, making her way to the comb being built to house new offspring and supplies. At the building site, a vertical landscape of hexagons is taking shape. Hundreds of other bees are already at work polishing bits of wax into smooth, ordered cells. The new arrival feels around, finds a spot in need of work, and waits for her material to be delivered.

It emerges, warm and liquid, from eight tiny slits between the overlapping armorial bands on her abdomen. Eight droplets of wax spurt to the surface of her body, where they cool into tiny, almost weightless flakes, ready for use. The fatty flecks are irregularly shaped ovals, little rounded scales bent at first to the curve of the bee's body. They are delicate and shiny, like mica or the husk of a kernel of popcorn. Most are no bigger than the head of a small pin. In the late eighteenth century, when the flakes were first discovered (it had been assumed previously that bees gathered wax from plants), a beekeeper named Dubini diligently took the time to try to weigh them. He determined that "These scales, of an irregular pentagonal shape, are so thin and light, that one hundred of them hardly weigh as much as a kernel of wheat."

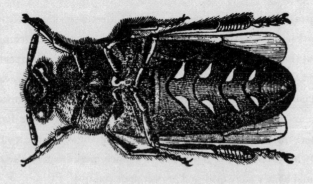

From *Langstroth on the Hive and the Honey Bee*, 1889.

The sheets of wax are left in their abdominal dispensers until the bee is ready to use them. First, she assesses the cell being built and finds a thin or ragged spot in need of more wax. Then, like a construction worker pulling nails from her toolbelt, she reaches for a flake with an available rear leg equipped with spiky tongs for this purpose. She pulls the flake of wax from her belt and in an advanced yoga move transfers it to her mouth. There, she masticates it, chewing and working the wax like a baker kneading dough, making it soft and malleable. She deposits this spongy, flaky load on the cell like a bit of mortar and then shapes, smoothes, and compacts it into polished comb. All around her, other masons are patting and caressing their own scales of wax into place. Each bee may spend only a few seconds actually shaping the flake, but the preparation of it—removing, softening, and then affixing it—can take as long as four minutes.

When a graft is done to her satisfaction, she wanders on to another building project, armed and ready with more abdominal flecks of wax. Soon, another bee will come along, flake in mouth,

One of the author's bees poses next to her waxen handiwork.

and make adjustments and additions to the work that her sister builder has just left. When a cell is brought to its full shape and height, a house bee will appear with a tongueful of drying nectar and load food into the new storage unit, or the queen might commandeer it to receive an egg. After the food cargo has been deposited, another shift of worker bees will bustle past, incrementally grafting wax lids to the top of the cell. One lid, the diameter of a lentil, can take several hours and dozens of bees to assemble.

The construction bee meanders around the site adding wax. Primping and polishing the cells as she goes, she is surrounded by hundreds of other workers doing the same. The bees work in clusters, keeping the temperature at the work site high and the building materials malleable. Flake by flake, nudge by nudge, they determinedly sculpt their home. Comb building is a chaotic communal effort, but the bees manage to make it appear effortlessly choreographed, as if there were a chore sheet or a song list hanging on the side of the hive. Lorenzo Langstroth, Maurice Maeterlinck, Charles Darwin, A. I. Root, and many others have been fascinated by the building of the comb and describe it as so incremental and infinitesimal that the cells appear almost as if by magic. Root writes, "The finished comb is the result of the united efforts of the moving, restless mass, and the great mystery is, that anything so wonderful can ever result at all, from such a mixed-up skipping about way of working as they seem to have.... The sum total of all these manoeuvres is that the comb seems almost to grow out of nothing." In the best conditions, the nothing happens fast. Ten thousand bees can produce one pound of beeswax in three days. Smiley has seen a mature hive create ten frames of wax comb, fill them with honey, and cap them off in just one week.

Without the help of man-made frames, foundations, or hives, bees will spin their waxen web from scratch. To furnish a wild tree hollow or an overturned bucket with walls and rooms, the bees grab on to one another's legs and hang down into the space in which they desire to build, forming a living plumb line that perfectly measures the comb and scaffolds the builders in place. Festoons of bees hang, like a pyramidal pep squad or a circus act, pulling wax from their bodies and sculpting it into slabs of flawless hexagonal comb.

An eighteenth-century interpretation of the bees' wax acrobatics.

In the wild or in a backyard box hive, eight pounds of nectar must be gorged by the bees to produce one pound of beeswax. This mass of wax can be sculpted into as many as 35,000 cells, or seven deep hive frames full of comb. Millions of weightless scales of wax and endless cementing pats produce both skeleton and skin of an amazingly sturdy organic structure. Though its walls are as thin, light, and translucent as tissue paper, the comb is able to support and contain, suspended, a weight thirty times its own. To me, the wax comb and its construction are as magical and fascinating as the honey they hold.

The precision, shape, and ingenuity of this architecture have fascinated beekeepers and scientists for centuries. Charles Darwin wrote: "It has been remarked that a skillful workman, with fitting tools and measures, would find it very difficult to make cells of wax of the true form, though this is effected by a crowd of bees working in a dark hive." To figure out how and why they create these endless hexagons joined together as neatly as a tile floor, Darwin offered his bees a test strip of smooth wax on which to build. He observed:

The bees instantly began to excavate minute circular pits in it; and as they deepened these little pits, they made them wider and wider until they were converted into shallow basins, appearing to the eye perfectly true or parts of a sphere, and of about the diameter of a cell. It was most interesting to observe that, wherever several bees had begun to excavate these basins near together, they had begun their work at such a distance from each other that by the time the basins had acquired the above stated width (i.e. about the width of an ordinary cell), the rims of the basins intersected or broke into each other. As soon as this occurred, the bees ceased to excavate, and began to build up flat walls of wax on the lines of intersection between the basins.

This choreography yields a meticulous hexagonal floor plan for the entire comb. The design is so uniform and precise that in 1792, when the First Republic of France was considering different systems of measurement, an eager beekeeper proposed that the cells of bees be used as a standard unit for distance. Luckily for the French, it was determined that not all cells are created exactly equal, and the idea was abandoned.

Squares or triangles would also accomplish the bees' goals of seamless coverage with no wasted space. The geometric genius of hexagons is that they require the smallest amount of precious building material to hold the greatest volume of honey. Charles Darwin determined that this shape was a brilliant evolutionary act. He wrote that the hexagonal cells

> *save much in labour and space, and in the materials of which they are constructed ... a prodigious quantity of fluid nectar must be collected and consumed by the bees in a hive for the secretion of the wax necessary for the construction of their combs. Moreover, many bees have to remain idle for many days during the process of secretion. A large store of honey is indispensable to support a large stock of bees during the winter; and the security of the hive is known mainly to depend on a large number of bees being supported. Hence the saving of wax by largely saving honey, and the time consumed in collecting the honey, must be an important element of success in any family of bees. Beyond this stage of perfection in architecture, natural selection could not lead; for the comb of the hive-bee, as far as we can see, is absolutely perfect in economising labour and wax.*

There are other perfections in the comb's construction. Each cell, for example, is angled slightly upward, about ten degrees from the horizontal, better to store runny nectar and keep it from seeping from the bee cupboards. The hexagonal shape provides five channels of ventilation in the brood cells as the eggs grow into plump rounded larvae. Wax is insoluble in water and nonconductive of heat, allowing the colony to fend off the elements and easily moderate temperatures within the hive.

Over the centuries, humans have found more uses for beeswax than its creators have. Until the invention of plastics in the late nineteenth century, there was no more plastic medium than wax. Inspired by the bees, we have used it for storage, building, and protection. Expanding on its versatility, we have used it in ceremony, as light, and for communication and art. In food storage, the earliest canning techniques were perfected by bees, who preserve their sugary sustenance in thousands of homemade wax containers that retain moisture, keep air out, and sturdily withstand the tests of time. Ancient Egyptians followed suit and enveloped their food in wax to protect and preserve it like honey in the hive. Wine casks and urns of olive oil and fruit were sealed with lids of it, and cheeses, meats, and eggs were dipped in it for storage.

Just as it coated and conserved a hunk of cheese, beeswax could be wrapped around a corpse to achieve the same ends. Once again, bees provided the example by entombing hive intruders in wax. Maeterlinck, in *The Life of the Bee*, writes:

> Bees will dispose of a mouse or a slug that may happen to have found its way into the hive. The intruder killed, they have to deal with the body, which will very soon poison their dwelling. If it be impossible for them to expel or dismember it, they will proceed methodically and hermetically to enclose it in a veritable sepulcher

of propolis and wax, which will tower fantastically above the ordinary monuments of the city.

Egyptians took lessons in embalming from the bees, wrapping bodies in wax-dipped linen bandages to preserve them for the afterlife. Ears and nostrils were often packed with wax and resins before the body and its removed organs were placed in a coffin or series of coffins sealed with beeswax. Herodotus, the Roman historian of the fifth century B.C., writes that many other early kingdoms wrapped their deceased royalty similarly, in cloths dipped in wax. Another chronicler wrote of the embalming of Agesilaus, the king of Sparta who died on the battlefield in 360 B.C.: "Having fallen ill, he died, and so that his friends could transport the body more easily back to Sparta…they enveloped his body in wax."

Sieves, screens, and presses were devised to help harvest the valuable preservative. Columella described the rendering process, which has not changed much since the first century:

> *The honeycombs, when they have been well squeezed, after being carefully washed in fresh water, are thrown in a brazen vessel; water is then added to them and they are melted over a fire. The wax is poured out and strained through straw or rushes. It is then… poured in such moulds as one has thought suitable.… When the wax is hardened, it is easy to take it out.*

Thus molded, bars, balls, and strips of wax were sold or traded in the markets of ancient Egypt, Rome, and Greece for a wide variety of applications, from embalming cheeses and bodies to polishing floors and walls. Like bees, humans used wax in building, preserving, and protecting their nests. Soon after it was taken from the hive, the fat of the bees was being rubbed into the walls, floors, furniture, and art of ancient abodes. Burnished with wax, surfaces of wood, plaster, clay, cloth, and leather were stronger, suppler, weatherproofed, and possessed of a luxurious shine. In Greece as early as the fifth century B.C., pigment was mixed with molten wax and then applied, still hot, to walls and surfaces, creating a rich and lustrous effect called encaustic, which means "burnt in." Twenty-five centuries later, molten colored beeswax is enjoying a renaissance on stylish walls and in contemporary artworks.

Two thousand years ago, the chief architectural scholar of the civilized world was a Greek named Vitruvius. In his *de Architectura*, or *On Architecture*, he described the use of wax as a fixative and wall polish for discriminating decorators:

> *Anybody who is more particular and who wants a polished finish of vermilion that will keep its proper color, should, after the wall has been polished and is dry, apply with a brush Pontic (purified) wax melted over a fire and mixed with a little oil; then, after this he should bring the wax to a sweat by warming it and the wall at close quarters with charcoal enclosed in an iron vessel; and finally he should smooth it all by rubbing it down with a wax candle and clean linen cloths, just as naked marble statues are treated ... the protecting coat of Pontic wax prevents the light of the moon and the rays of the sun from licking up and drawing the color out of such polished finishing.*

The waterproof luster of wax also made it a favorite covering for boats. The Greek historian Callisthenes described a famous fighting ship on which "every part was decorated with wax painting." Homer also writes of the painted black prows of the Greek warriors at Troy. Wax brought aesthetic elegance, durability, and waterproofing to almost any surface. Even fresh fruit could be kept longer and made more attractive with a shiny coat of wax. In the Middle Ages, leather armor that had been boiled for toughness was afterward rubbed with wax to buff, preserve, and waterproof it, and also to imbue it with the blessing of the courageous bees. In later centuries, circus and camping tents, raincoats, and tablecloths were all commonly sealed and strengthened with beeswax. It seems that human lives throughout history would have been wet, rotten, and dull without bees and their wax.

In the ancient world, beeswax soon progressed from preserving boats and bodies to preserving likenesses. It was more versatile than wood, clay, or even gold, and could capture exact, minute impressions. Egyptians placed so much emphasis on the body in the afterlife that, as insurance, they fashioned miniature wax replicas of the deceased in case something happened to the original. Placed in their own little coffins and inscribed with identical hieroglyphs, the wax figures could take the spiritual place of the body if necessary. Replicas of favored gods were also made out of wax and used for prayer or entombed with the body for good luck in the afterlife. Religion and beeswax were entwined in life, in death, and beyond.

Important objects of personal and spiritual use were often made from the loss of wax rather than the conglomeration of it. For centuries, artists in Mesopotamia, the Mediterranean, Africa, and China have crafted treasures using a method known as *cire perdu*, or "lost wax." First, the desired object was precisely modeled or sculpted in wax, then coated with clay to form an exact, detailed mold that

Wax icons found in a tomb in Luxor, Egypt, 1500 B.C.

hardened around the core. When the whole was fired, the wax was "lost" as it melted out, leaving a baked clay mold into which molten metal could be poured. Without beeswax and its heated loss we would not have goblets and icons from the temples of ancient Sumeria or golden amulets from Egyptian tombs. Waxless, the city of Florence would be without its Cellinis, and the Forbidden City in Beijing would not have imposing bronze dragons guarding its gates.

Death rituals and elaborate funerals were a frequent, significant occurrence in ancient cultures, and wax played an important role. Detailed likenesses of the dead, even before they deceased, were necessary, auspicious, and impressive, and wax artisans were kept busy making face masks, portraits, and busts out of colored, complexly worked beeswax. These likenesses of living and dead family members were proudly displayed in the atria of fashionable Greek homes. In Roman funeral processions, flattering wax masks of the dead were adhered to coffins for exhibition through the streets, with an encaustic portrait of the departed often leading the parade.

Because of bees' connection to the deities, their wax also enjoyed a special status. Karl von Leoprechting wrote, in 1855, "The bee is the only creature which has come to us unchanged from paradise, therefore she gathers the wax for sacred services." Ancient cultures used this blessed substance in a variety of religious practices, and offerings of wax and honey were common. In the households of Greece and Egypt, icons, charms, amulets and votive figures were modeled in wax and relied upon for godly favors and protection. Sickness or injury could be cured by creating a wax model of the afflicted organ or limb and praying for it or treating it in effigy. In parts of South America and Europe, these miniature body parts and waxen wishes can still be purchased near churches and hospitals.

Beeswax was also employed in the darker pursuits of sorcery and witchcraft. In ancient Greece, wax symbols and effigies were placed on houses or at roadsides to invite a hex on any or some who passed. Plato wrote of the trepidation associated with them: "And when men are disturbed in their minds at the sight of waxen images, fixed either at their doors, or in a place where three ways meet, or on the sepulchers of their parents, there is no use in trying to persuade them that they should despise all such things because they have no certain knowledge of them." Black magicians created wax effigies of the victim of a spell and then visited punishment upon them in the form of pins, heat, cold, dismemberment, or mutilation, depending on the desired result. Spells could be worked on individuals or on events with equally devastating results. King

Nectanebus, a warrior who ruled in Egypt in 350 B.C., used beeswax to thwart his enemies. Fashioning a minifleet of ships from wax, he positioned them in a bowl of water. After invoking a spell, it is said, he watched gleefully as his miniature nemesis sank to the bottom of the bowl. His real enemies soon followed suit and drowned at sea.

In 1315 in France, a man named Enguerrand de Marigny, his wife, and his sister-in-law were all hanged for the crime of "image magic" against Louis X and his uncle Charles of Valois. They had crafted waxen figures representing the royals, stuck them with pins, and placed them near a fire, believing that as the icons gradually melted away, so would the power of their victims. (I've tried this, but I must have done something wrong, as my targets are still in office.) Charles was racked with stabbing pains and near death when the plot was discovered, but as soon as the effigies were destroyed, he recovered, quickly enough to attend the hanging.

Good, evil, and art consumed tons of wax. What was left over was applied to more mundane and practical purposes. Beeswax was man's first plastic, the versatile superglue of its day—used for all kinds of mending, building, and binding. The god Pan, making an early flute, "taught to join with wax unequal reeds." Mortals routinely joined false teeth together using the same material. Daedalus used beeswax to make the wings with which to fly off the island of Crete—forgetting as he flew close to the sun that this fabulous fix-all melted easily. In the *Odyssey*, Ulysses used beeswax to model plugs that would keep the alluring call of the sirens from the ears of his men. "I with my sharp sword cut into small bits a great round cake of wax, and kneaded it with my hands, and soon the wax grew warm, forced by the strong pressure and the rays of the sun. Then I anointed with this the ears of all my comrades in turn."

Contemporaries and descendants of Homer used wax as an all-purpose adhesive, to hold the tesserae of mosaics in place, and to mend cracks in pottery. It was also used for correspondence, before papyrus became widely used. Tablets made from wood were covered with a thin, smooth coat of wax that served as a versatile, portable, and reusable writing surface. Characters were etched into the sheen of wax using an instrument called a style, which had a pointed end for writing and a spatulate end for smoothing and erasing. Wax tablets and books spread from Egypt to Greece and Rome, and continued to be used well into the fourth century. Catullus, a poet of the first century B.C., extolled the perfection of wax tablets for sending secrets and love letters, as once the contents had been read, they could be erased and a reply immediately returned, scratched onto the same receptive surface.

The earliest humans learned to burn some kind of fat to produce light, using resinous woods, animal tallow, and even, in some parts of the world, strips of oily fish. The first torches were splinters of fatty woods that were sometimes dipped in additional lard or coated with beeswax. Over the centuries, wax coatings got thicker, the splintery "wicks" got thinner, and by the turn of the first millennium, candles as we know them were born. Catullus might have used the light from one to read his waxen love letters.

From the Acropolis in Greece to the cathedrals of medieval Europe, compact, portable beeswax candles with their self-sufficient sources of fuel were the most desirable source of light. They burned so cleanly and reliably that they were even used to measure time. Beeswax also lit easily, burned brightly, and left no objectionable odor or murky residue, as did a smoldering strip of salmon or fused animal tallow. "Base and unlustrous as the smoky light that's fed with stinking tallow" is how Shakespeare described the effects of the poor man's wax alternative. Light from a beeswax candle was the lustrous, odorless best—and the most expensive.

From the tenth century until the sixteenth, the church was the biggest and wealthiest consumer of wax in the world, and bees were put to work as much for their combustible wax as their consumable honey. Because of their purity and powerful religious symbolism, beeswax candles were in high demand and mandated for ceremonies of all kinds. In the fifteenth century, one parish in Wittenberg burned twelve tons of wax every year. In 1422 the funeral parade of Henry V stretched over two miles long and was brightly illuminated with candles. When the cortege reached Black-heath, it was met by a delegation carrying 1,400 glowing wax tapers. Elizabeth I outshone Henry V even in death. At her funeral, 10,000 torches lit the procession, and 500 candlesticks illuminated the bier. Given the church's taste for dazzling pomp, there was not always enough beeswax to supply the demand, and a busy international trade sprang up to meet the need, with shipments coming from as far away as outsourced bees in Africa and China. This commerce abated in the sixteenth century, when the dark stringencies of the Reformation curtailed the need for candles.

By the twelfth century in Europe, with the high value of wax and the voracious demands of the church, candle making had become a specialized craft and enviable business. The candler, or "chandler," was part of an official guild, whose members were the purveyors, fabricators, and inspectors of fine wax candles. Because of beeswax's value and opacity, there was great temptation to corrupt and dilute it to increase profits. In Morocco, heavy stones were often dropped into exported bricks of wax, which were purchased by weight, and in Europe, animal fat was added. One of the chandler's responsibilities was to guarantee the provenance and quality of the wax. In 1580 Queen Elizabeth, who had a passion for pure bees wax candles, passed an "Acte for the true melting making and working of waxe," which harshly penalized stone and fat adders and required merchants to stamp, emboss, or mark their product with initials and guarantees of purity.

Beeswax illumination was for royalty, the church, and the very rich. In 1432 a pound of bleached white wax, the kind used in church candles, cost as much as a typical laborer could earn in a day. On the other side of town, the housekeeping accounts of an English earl show he consumed 1,870 pounds of wax for candles and 1,700 pounds of wax for seals (which signed and authenticated documents and correspondence) in one year. Throughout Europe, aristocrats used wax and light to pay their employees. High-ranking household staff were often salaried in cash, food, wine, and candles. Henry I's chancellor, for example, received a salary of five shillings a day, along with wine, bread, and "l fat wax candle and xl pieces of candle." At all levels of society, rents, fines, taxes, bribes, gifts, and goods could be paid for in wax. Some villages even stamped coins out of it and used them in trade. In 1513, the English traveler John Hoy journeyed to North Africa and was appalled to find that the residents of a town called Hadecchis had no use for the commodity, which was as good as gold in his own country. "Because they know not what service to put their wax unto, they cast it forth, together with the other excrements of honey. No people under heaven can be more wicked, treacherous or lewd…than this people is."

From its ancient beginnings in religion and death, wax model-ing and portraiture matured into a popular art form. Craftsmen were enamored with the infinite plasticity and delicacy of wax, adding pigments to create a complex palette, then rolling, pinch-ing, and scraping it to model extraordinary replicas of life. The first-century Roman historian Varro praises the skill of a wax worker who fashioned "apples and bunches of grapes so like nature that it was impossible to distinguish any difference at sight between the real and artificial." This tradition began in Rome and persisted, with a medieval hiatus, into the nineteenth century, where wax modeling was considered an enviable parlor skill. A Boston news-paper advertisement from 1881 reads, "Wax Work. Mrs. A. D. Forrest. Manufacture and designer in all kinds of wax work. Duplicating and pressing flowers. Orders taken and lessons given in wax fruit, flowers and confectionery. Also in preparing wax for flowers copy-ing from natural plants, blossoms, and autumn leaves. Wax figures made to order." Countless magazines, newspapers, and books offered instruction in the delicate art of imitating life in wax, catering to the fascination with both botany and illusion at the end of the nineteenth century. Auguste Escoffier, the great chef, restaura-teur, and style maker, published a best-selling book called *Flowers in Wax* in 1910, teaching readers to transform beeswax into ferns, flowers, and fruit.

In portraiture, wax artists rivaled the best painters and sculp-tors of their day. Wax captured color and detail in three dimensions and was in many cases the preferred medium for preserving a like-ness. Giorgio Vasari, in *On Technique*, marvels at the abilities of wax modelers in 1550s Florence: "Modern artists have discovered the method of working in wax of all sorts of colors, so that taking portraits from life in the half relief, they make the flesh tints, the hair, the clothes, and all other details so lifelike that to those figures there lacks nothing, as it were, but the spirit and the power of speech." When Queen Elizabeth died in 1603, expert modelers went to work. A resplendent figure, made of lifelike wax (though slightly younger and fresher looking than the queen's seventy years), was dressed in funeral finery and carried in procession to Westminster Abbey.

One hundred and fifty years later, a young Frenchwoman named Marie Grosholtz took up the art of wax portraiture. As a girl, she had been employed by a physician who was a skilled wax worker (doctors and scientists often used wax to create anatomical likenesses and indeed entire educational corpses). Grosholtz's exquisite modeling skills took her to the French court at Versailles, where she fashioned portraits of royalty and a constellation of courtiers. When revolution broke out, she went to the guillotine to take death masks of many of these same faces (now without bodies), including those of Marie Antoinette and Louis XVI. Grosholtz married a Frenchman named Tussaud, and, as the turmoil in France continued, left for Britain to tour her work, which was by this time an impressive collection of deceased people. Madame Tussaud became an instant sensation with her chillingly real wax likenesses, as most people had never seen the celebrities of the day, dead or alive.

The craft and reportage of wax modeling thrived for another hundred years in most of Europe and America, until more accurate means of capturing likenesses arrived. E. J. Pyke, in the introduction to his *Biographical Dictionary of Wax Modellers* (which chronicles the superstars of wax modeling in Europe and America) wrote, "The end of the nineteenth century saw, to all intents and purposes, the disappearance of wax modeling with the advent of photography." Sugar, the bane of honey's existence, also had a hand in the disappearance. The popularity and price of sugar extinguished the backyard apiary, drastically reducing the production of wax for modeling and other arts.

Before its demise in portraiture, wax was migrating into other creative areas. In Bavaria in 1798, a frustrated writer and performer named Alois Senefelder was experimenting with printing and etching techniques in hopes of finding a more lucrative way to make a living. The story goes that he was in his home studio when his mother asked him to write down a grocery list. Having no paper at hand, he grabbed a nearby hunk of beeswax and wrote the list on an etching stone. Perhaps because he didn't want to lug the heavy limestone to the market, he started experimenting with how to get the wax list off the stone. Moistening it, he noticed that water didn't mix with the oily beeswax and that pigment rolled onto the wax would also repel the water. When paper was pressed to the surface of the stone, the beeswax grocery list came off in perfect reverse, and lithography, or "stone writing," was born.

Prior to the twentieth century, beeswax and other oil crayons were of the kind Senefelder knew, pigment-free and used mostly for commercial lithography and other industrial purposes. At the turn of the century, Edwin Binney and his nephew, C. Harold Smith, who were in the paint business, thought there might be a market for colored wax sticks and began experimenting with beeswax and some of the newer petroleum-based varieties. In 1903, they produced the first rainbow box of eight wax crayons, which they sold successfully to schools. Alice Binney, Edwin's wife, christened them "Crayolas" by joining the French word *craie*, or chalk, with "ola," short for "oleaginous," or oily.

Many of the world's great creations have a trace of beeswax somewhere in their history. While Binney and Smith were experimenting with wax and color, Thomas Edison was testing wax and sound. Edison's phonograph was first developed using a steel needle and tinfoil to capture the audio impressions of his voice. Tin tore easily and produced a muffled recording, so Edison turned to beeswax, aware of its ability to capture detailed impressions. He substituted a wax cylinder for the tinfoil and recorded tiny scratches and grooves of sound. Applying the ancient technique of lost wax, he applied a micro-thin layer of gold atop the wax so that heavier layers of metal could then be added to create a mold. When the wax was lost and vinyl was added to the mold, the permanent record of sound was gained.

Wax Craft, a 1908 book, includes more than one hundred detailed, practical recipes for beeswax. Instructions for waterproofing leather, removing stains from marble, and making bottles airtight are included, as are recipes for lipstick, sealing wax, boot blacking, waxed paper, sizing for linen, and sealants for the cracked hooves of a horse. Readers learn how to whip up modeling wax, crayons, shoe polish, furniture polish, varnish, floor polish, and shaving paste with beeswax. The turn-of-the-century home, like the ancient Greek one, would clearly have been rough and ungroomed without it. Modern householders can stroll through a supermarket and still find the ancient sheen, polish, and protection of beeswax in furniture and shoe polish, crayons, candles, lip balms, lipsticks, and hand creams. Cosmetics companies are the biggest users of beeswax, followed by candle makers and beekeeping suppliers. Pharmaceutics and dentistry also use quantities of it, as do makers of floor polish, automobile wax, and chewing gum. In his book *Adventures in Man's*

First Plastic: The Romance of Natural Waxes, published in 1947, Nelson Knaggs writes: "Chemists were quick to pick up where the ancients left off and exploit the qualities of waxes. Our floors, shoes, furniture, automobiles, boats, and aircraft are polished incessantly with a coating of wax. America virtually walks, rides, floats and flies on a film of wax."

Though many beekeepers sell their wax to distributors or candle makers, Smiley puts all of his back into his business. Soon after he returns from Dadant, he spends a morning assembling the new frames for which he traded his wax bricks. Into each he slides a sheet of wax foundation from Dadant, the recycled work of thousands of fatted bees. With a contraption he's rigged up using wooden thread spools as pulleys and a fishing reel full of wire, he threads a zigzag of metal filament across the foundation as if he were weaving on a miniature loom. Reeling in, he stretches the wire taut across the sheet of wax, strumming it like a guitar to test it. Nodding, he is satisfied that the frame is strong enough for the weight of the next honey season and the force of the extractor. Threading and strumming, he builds a pile of reinforced frames. These will be put into hives to replace those grown old with weather and wear. As soon as the bees get their feet on these new walls, they'll start building cells and cities out of tiny flakes, surrounding themselves, as we humans do, with a vital shiny film of wax.

Smiley meanwhile surrounds himself with bills and accounting work, the part of the job he likes the least. "I feel so lost in this paperwork sometimes," he moans, wishing he were lost in a busy beeyard or out fishing on the river. At least he has a nice new office in which to struggle. The Smileys finished moving into the pink house in mid-October, carrying stuff over by hand and using the truck and trailer for some of the bigger items. His heavy old metal desk now rests on a sea of brownish carpet with a wavy green fern pattern. The wall color, which he picked himself, is a soothing cocoa. "This is my favorite room in the house," he says, admiring it and waving at his growing collection of bee kitsch and paraphernalia. His office display includes cookie jars, plant potters, and picture frames, all decorated with bees. There is a bee paperweight on his desk and a teddy bear with an acorn-sized plush bee on its nose. Bee magnets cling to his metal filing cabinet, and even the clock over the arched doorway has hives, leaves, and bees at the hour markers. "I'm collecting some nice stuff," he says proudly.

Seated at his desk amidst this collection, he enters expenses, labor costs, and sales in pencil every month in a tall cardboard-bound ledger book. By the end of October, with the last crops of wax and honey sold and tallied, he usually knows how he did for the year. He harvested a total of 122 barrels of honey. Foreign honey prices and rain cut into his profits a bit, but he's happy. He caught and sold 50 barrels of tupelo. "I did just fine," he says. "I'm sitting pretty." Prettier now that the hassle of building is over and the house is complete. "I know some other beekeepers who built houses and their businesses went down," he says. "But my business was good. I just refuse to neglect my bees." He looks around. "But if I ever did it again I'd get a contractor." Before he dives into the dreaded paperwork, he procrastinates with one of his favorite activities: thinking about bees and business expansion. "I'm going to have a thousand hives next year," he says, a frequent refrain. For a moment, he fantasizes about that rich man's honey house he'll build; then he picks up his pencil and goes to work.

NINE

Medicine Ball

The Lord hath created medicines out of the earth.
Ecclesiasticus, 38:4

*If the dew is warmed by the rays of the sun, not honey
but drugs are produced, heavenly gifts for the eyes,
for ulcers and the internal organs.*
Pliny the Elder, *Natural History*

*There proceedeth from their bellies a liquor of
various color, wherein is medicine for man.*
Qu'ran, 16:69 on bees

In winter, many people inaccurately suppose, bees and their keepers sleep a lot. In fact, in November and December Smiley is busily preparing for spring. His livestock is still up north, eagerly anticipating the same. There is not much to forage on anymore, so the bees feed on stored-up honey and look forward to the first awakening scents of spring. The cotton fields near where the bees are bivouacked are now acres of dark brown, prickly stalks, naked and withered except for a few stray wisps of bedraggled cotton hanging like toilet paper from trees on the day after Halloween. In lots where the plants have been trimmed to the ground, stray cotton collects in dirty tapered drifts like melting snow.

It rarely snows in the panhandle (Smiley can remember three storms in his lifetime), but it regularly gets cold enough to freeze above-ground pipes and send bees into their winter formation. They cluster loosely and casually all year, but when the temperature drops below fifty-seven degrees, they contract into a basketball-sized scrum within the hive. This tight cluster is the winter furnace for the colony, fueled by honey and the intentional shivering of the bees as they contract their wing muscles. On the surface of the ball, workers create a thick, insulating shell of live bodies that hovers at around forty-five degrees no matter how low the temperature drops. The bees forming the crust are inert, conserving their energy and accepting the heat generated by the bees within. The interior grows progressively warmer and busier toward the core, where the queen goes about her reproductive business at a constant balmy temperature of around ninety-three degrees (even if there is snow drifted on top of her hive). During long cold periods, the bee ball with its royal core moves slowly around the nest, consuming the food it has stored. Bees don't sleep or hibernate in winter, but, like many of us, they become generally lethargic and subdued, resting and preparing themselves for the warm rigors of spring.

In the coldest months, when nectar and hive activity are at an annual low, the queen herself usually takes a holiday. She doesn't actually get to leave the workplace but does quit her laying job for about four weeks, usually in December. In January, refreshed, she goes back to work building up the hive population for spring. Worker bees also get a reprieve of sorts in winter; they are spared the mileage and wear of forage and live up to four times longer than their summer counterparts. The monarch and her elderly crew all slow down in the cold. Their groggy mandate is to eat, stay warm, keep queen and brood comfortable, and be prepared to charge out of the hive as soon as they sense the return of nectar.

In winter, while the bees are chilling, the first nectar flows of the Florida panhandle are only a couple of months away, and Smiley has to repair old hives and build new ones to catch them. A beekeeper's winter looks a lot like a carpenter's. The shed and yard next to Smiley's honey house look like an outdoor furniture factory, with drifts of sawdust on the hard-packed ground beneath an assortment of tools, paint cans, table saws, and stacks of raw pine lumber. Rolling up his long sleeves and taking a swig of coffee, Smiley flips on a saw, grabs a long board, and goes to work. Over the course of the next several weeks, between drive-bys, he and George will transform the pile of lumber into 400 new hives to house 400 new colonies of bees in the spring.

There are many ways to get new bee colonies. Swarms can be trapped in the wild (although those numbers are dwindling), ordered from a supplier, or raised in the apiary. Experienced beekeepers have many different methods of making new livestock. All involve taking brood and food frames from existing hives, borrowing some workers to service them, and adding a new queen they've purchased or raised to mother and multiply them. It's like taking bread dough from different loaves, adding some royal yeast, and creating an additional bustling loaf of bees. This beekeeping baking is done all the time, but most often in early spring, so that the new colonies will be kneaded and risen, ready to chase nectar by April and May. Smiley makes boxes all winter to house the new spring loaves, called splits or increases. The boxes he builds this winter, and the bees he'll raise in them in the spring, will increase his annual honey production by about 50,000 pounds if all goes well.

While Smiley cuts pine board into box-size lengths and uses another saw to incise carrying handles deep into them, George is nearby, at work replacing splintery wooden frame parts that were bitten and chewed in the metal teeth of the uncapping machine. In the fall construction work, there are often minor beekeeping injuries. "Me and George are always getting banged up around here," Smiley says, inspecting a bloody scratch on his arm. Fortunately, he has a limitless supply of one of the world's great healers at hand and can smooth honey onto most of his minor scrapes. For as long as humans have been acquainted with bees, they have been getting medicine from them in the form of honey, propolis, pollen, and even venom. The hive is an ancient apothecary, loaded with convenient, economical, and powerful remedies.

The antibacterial and moisturizing qualities that made honey invaluable in the kitchen also made it a staple of the doctor's bag. Its hydrogen peroxide, which helped preserve meats and fruit, was also an effective cleansing treatment for wounds, aided by the osmotic thirst of honey's sugars. When applied to an infection, the absorbent sugars in honey act as healing sponges, draining intruding organisms of their liquid essence and causing them to shrivel and die. At the same time, the sugars nourish healthy cells and encourage white blood cells in their healing battles. Antioxidants, amino acids, and vitamins in the natural ointment reduce inflammation and speed the growth of healthy tissue. A golden smear of honey across a splinter-pricked thumb or scratched arm cleans the wound and promotes healing while providing a moist protective barrier. As he saws and scrapes, Smiley often employs the world's first self-adhering Band-Aid, complete with its own vitamin-rich antibacterial ointment.

Smiley applies his product to cuts and burns and uses it in a sore throat gargle. Using his mother's recipe, he combines about two tablespoons of honey, an eighth of a teaspoon of alum (an astringent he gets from the grocery store), and a teaspoon of lemon juice in a pint jar. "I fill that jar with water and keep it in the fridge and gargle with it. Boy, that'll knock the sore throat right outta you," he says. "Make your mouth feel like you're chewing on a cotton ball, but it works. I guaran-damn-tee it."

Ancient Egyptians had a similar guaran-damn-tee'd recipe for sore throats. The oldest known intact book of medicine is the *Papyrus Ebers* (named after the German Egyptologist who acquired it in 1873), a fifteenth-century B.C. collection of more than 800 medical problems, diagnoses, and prescriptive recipes, more than half of which employ honey. Used alone or combined with animal, vegetable, and mineral compounds, the produce of the hive was used to treat everything from head wounds to human bites. For an aching throat, the *Papyrus Ebers* suggested a bit of goose grease combined with honey, incense, caraway seed, and some water, rolled into a ball, and chewed nine times. To heal an eye, "when something evil has happened to it," the papyrus calls for a human brain and some honey. "Divide it in halves. To one half add Honey and anoint the eye therewith in the evening. Dry the other half, crush, powder, and anoint the eye therewith in the morning." To induce an abortion, or "for loosening a child in the belly of a woman, one part fresh beans and one part honey, pressed together and drunk for one day."

Each of these remedies would have been administered by a priest, who would strengthen its effect by chanting spells and incantations as he applied his honeyed potions. Illnesses at this time were thought to be the manifest displeasure of the gods, so remedies had to both relieve symptoms and placate the deities. Physicians were priests, sorcerers, and shamans combined, and honey, with its supernatural support, was a persuasive, popular healing staple.

Classical Greece inherited much of its healing herbal repertoire from the Egyptians, the acknowledged experts. In the *Odyssey*, Egypt in 800 B.C. is described as the place where "There grow all sorts of herbs, some good to put into the mixing-bowl, and others poisonous. Moreover, every one in the whole country is a skilled physician." Greeks celebrated and expanded the Egyptian pharmacopoeia, though they were less enthusiastic about the spells and sorcery. In classical Greece, illness was not a vengeful act of God but a bodily imbalance that could be remedied with knowledge. By the fourth century B.C. the Greek physician Hippocrates and his followers had catalogued the ailments of the world, enumerating over three hundred drugs and their uses while pointedly omitting magic and superstition. For the treatment of hemorrhoids, for example, Hippocrates offers advice free of hocus-pocus and full of brutal, honeyed practicality.

Have the person lie on his back, and place a pillow beneath the loins. Force the anus out as far as possible with your fingers; heat the irons red-hot, and burn until you so dry the hemorrhoids out that you do not need to anoint: burn them off completely, leaving nothing uncauterized. Let the assistants hold the patient down by his head and arms while he is being cauterized so that he does not move.... After you have applied the cautery, boil lentils and chick peas in water, pound them smooth, and apply this as a plaster for five or six days. On the seventh day, cut a soft sponge as thin as possible, place a piece of fine thin linen cloth equal in size to the sponge on top of it, and smear with honey [to make a poultice for the wounds].

Oozing ulcers seem to have been a prevalent problem in classical Greece. Hippocrates suggests that a sufferer "place lumps of very dry salt in a small copper vessel or a new pottery one, and add to the salt the finest honey, estimating the amount to be twice that of the salt. Then place the little pot on the coals, and leave it there until everything is completely burnt; then sponge the lesion off clean, and attach the application with a bandage." Honey was the enlightened doctor's versatile salve and tonic, used throughout the Hippocratic medical repertoire.

Hot irons and salt rubs could be avoided altogether (and this sounds like a good idea) if one lived a balanced life with a healthy diet, a bit of exercise, and a lot of honey. "Wine and Honey are held to be the best things for human beings," Hippocrates wrote, "so long as they are administered appropriately and with moderation to both the well and the sick in accordance with their constitution; they are beneficial both alone and mixed." It is said that the philosopher Pythagoras ate a strict vegetarian diet rich in honey and lived to the age of 100. Democritus, the renowned physicist, is also believed to have lived for almost a century, perhaps double or triple the average of the day, and divulged his secret as "oil externally and honey internally."

Chinese medical use of honey was similarly practical, reverential, and broad. Shen Nong's *Book of Herbs* from the second century B.C. extolled honey as a top-grade healer that could be taken for specific ailments and general good health. "It helps to kill pain and relieves internal heat and fever and is useful to many diseases. It may be mixed with many herbal medicines. If taken regularly, one's memory may be improved, good health ensues, and one may feel neither too hungry nor decrepit." The first ayurvedic doctors in India admired and employed honey in many of the same ways. In Mecca, the prophet Mohammed said, "Honey is a remedy for every illness."

Three centuries after Shen Nong's *Book of Herbs*, Pliny the Elder was at work compiling his massive thirty-seven-volume *Natural History*. He devoted fourteen of the volumes to cataloging plants, animals, minerals, and their medical uses, concluding that "There is nothing which cannot be achieved by the power of plants." He might have added, "and honey," as almost every one of the healing plants he described was paired with honey to effect various cures. Olive leaves pounded with honey could be used for "suppurations and superficial abscesses." Almonds and pomegranate rind combined with honey "is good for the ears, kills the worms in them, and clears away hardness of hearing, vague noises and singing, incidentally relieving headaches and pains in the eyes."

Sometimes, the power of plants was mixed right into the honey cure. Bees, in their divine wisdom, are strongly attracted to medicinal plants. Like flying herbalists or perfumers, they select and mix plant essences into their healing food. Pliny was aware of these essences and suggested different honeys for particular ailments. Attic honey, for example, from the thyme-blanketed hillsides of

Athens, was thought superior for treating eye problems, infused with medicinal properties whose fame was widespread. In the *Papyrus of Zenon*, Dromon writes to Zenon: "Order one of your people to buy me a kotyle of Attic honey, for I need it for my eyes, by command of God." Pliny's contemporary, the great physician Dioscorides, noted that "Honey made in Sardinia is bitter, because of the food of wormewoode, yet it is good for sunne burnings and spots on the face, being anointed on."

Dioscorides compiled *De Materia Medica*, a colossal five-volume book of medicine, while traveling the world as a doctor in Nero's army. Throughout his journeys, he assembled the first comprehensive universal medical dictionary, listing over 1,000 drugs and their uses. The first of several enthusiastic pages on honey or "mel" begins:

> *Attick honey is the best and of this that which is called Mymettium, afterward that of the cyclad islands and that which comes from Sicilie, called Siblium. That is best liked of, that is sweetest, and sharpe, of a fragrant smell, of a pale yellow, not liquid, but glutinous and firme when in ye drawing doth leap back to ye finger. It hath an abstersive faculty, opening the pores, drawing out of the humors, whence being infused it is good for all rotten and hollow ulcers. Being boyled and applied it conglutinates bodies that stand asunder one from the other, and being good with liquid Allum and so applied as also the noyse in the eares and ye paines, being dropt in luke-warme with sal fossils beaten small and being anointed on, it kills lice and nitts. It also cleanseth the things that darken the pupillae oculi. And it doth also heale inflammations about the throate, and about ye tonsillae, and the squinsies, being either*

anointed on or gargerized; it doth also move urine and cures coughs
and such as are bitten of serpents.

In addition to having its own entry, Mel is found throughout *De Materia Medica* under uses for various herbs. Of rue (a mountain herb), for example, Dioscorides asserts: "Ye juice being anointed on with ye juice of fennel and honey it helps ye dullness of sight." (This is first-century medicine; please do not try at home.) Of lentils, that Hippocratic hemorrhoidal favorite, he writes, "with honey doth conglutinate the hollowness of sores, it breaks scabs of ulcers round about, and cleanses the ulcers." Of alum roots (Donald Smiley's sore throat remedy), "They close moist gums, and with vinegar or honey they strengthen wagging teeth." With its comprehensive nature, clinical calm, and abundance of honey, Dioscorides's *De Materia Medica* was the *Physician's Desk Reference* of its day and the standard medical reference for the next seventeen centuries. Even today, while there are better treatments for snake bite and wagging teeth, many of Dioscorides's honeyed remedies, such as gargerizing the tonsillae (gargling for a sore throat and tonsils) or anointing it on sores, help soothe and heal.

Animal, mineral, and herbal cures were translated back and forth between cultures and languages throughout the Middle Ages, but their classical content did not change. Even the great tenth- and eleventh-century physicians of Arabia borrowed from Dioscorides, as did the healers in Europe. Hildegard of Bingen, a twelfth-century German abbess renowned for her healing abilities, compiled *Physica*, a book of remedies typical of the day and obviously indebted to Dioscorides. For a person who has "turbulent eyes," she writes, "so that at times there is a cloud which fogs them in some way, should take the sap of rue, and twice as much pure liquid honey, and mix them with good clear wine. He should put a crumb of whole wheat bread in it, and tie it over his eyes, at night, with a cloth."

Herbal remedies were honey-sweetened heirlooms, handed down from generation to generation, always tracing their ancestry back to *Ebers* and Dioscorides. A colic remedy from *The Queen's Closet Opened* of 1655 is not much different from the curing plasters of ancient Egypt or Rome. A Persian manual of medicine from 1685 lists an alphabet of diseases, from apoplexy to worms, and then offers a range of honeyed herbal treatments. If one's ailment was impotence, for example, there were a couple of sweet variations on Viagra. "For erection: the seed of the herb-rocket, and pepper with honey morning and night take as much as you can take up to two fingers. But if a man be grown old, and have a loose and hanging member, he shall do this. Of seed of rocket, cumin, pepper, and seed of purslane, being bruised and made up with honey, let him take it morning and evening. It is incomparable."

During the nineteenth century, honey was still utilized in remedies as it had been in the temples of Egypt and the apothecaries of Greece and Rome. Lorenzo Langstroth reported that

> ... *in Denmark and Hanover, the treatment of Chlorosis [anemia], by honey, is popular. The pale girls of the cities are sent to the country, to take exercise and eat honey.... Honey, mixed with flour, is used to cover boils, bruises, burns etc. It keeps them from contact with the air, and helps the healing. Beverages, sweetened with honey, will cure sore throat, coughs, and will stop the development of diphtheria.*

Doctor John Monroe of Vermont wrote *The American Botanist and Family Physician* in 1824. Echoing the enthusiasm and breadth of Dioscorides, he summarizes honey as "detergent, expectorant, emollient, demulcent, and highly purgative. It powerfully promotes expectoration; deterges and resolves rigidities…temperates that acrimony of the humors; helps coughs, asthmas, disorders of the kidneys, and urinary passages, and the sore mouth and throat. It closes ulcers, purges moderately, and resists putrefaction."

Today honey is still widely used as an emollient, detergent, and reliable resistor of putrefaction. This is especially so in poorer countries, where it is cheap and where traditional herbal healing is practiced. Kolawole Komolafe wrote recently in *Medicinal Values of Honey in Nigeria* that most of the 200,000 healers in his country include honey in their preparations. In his 1990 book on health sciences in India, B. L. Raina wrote, "In all parts of India, the elderly ladies in the house or neighborhood often suggest treatment of common complaints. The articles suggested for treatment are found in the kitchen, such as honey and ginger, or in the backyard, or readily available with the grocer, or growing wild in the neighborhood."

In "progressive" Western medicine, honey and other folk remedies fell out of favor during World War II as ambitious chemical drugs were developed. Lately, however, a medical honey renaissance has occurred as trials on animals and humans have validated the natural ointment. The journal *Burns* published a report in 1998 comparing honey treatment on burn victims to the standard of silver sulfadiazine. Of the 52 patients doctored with the natural salve, 87 percent healed within 15 days, compared to 10 percent of those soothed with chemicals. Honey recipients also experienced less pain, wound leakage, and scarring. Peter Molan, a doctor who codirects the honey research unit at Waikato University in New Zealand, has said, "Randomized trials have shown that honey is more effective in controlling infection in burns than silver sulphadiazine, the antibacterial ointment most widely used on burns in hospitals.... At present, people are turning to honey when nothing else works. But there are very good grounds for using honey as a therapeutic agent of first choice." Topical honey treatment is not a first choice just for burns. Doctors in India who applied honey and taped the outer incisions after a caesarean section found it to be more effective and less painful for the patient than chemical dressings and sutures. German doctors used a mixture of honey and anesthetic to treat herpes zoster, or shingles, successfully. There are many, many examples from all over the world of the ancient healer soothing a variety of modern wounds and complaints.

Taken internally, honey has been found to kill *Heliobacter pylori*, the bacterium that causes ulcers, and is even making inroads on staphylococcus infections. In *Honey, Mud, Maggots, and Other Medical Marvels: The Science Behind Folk Remedies and Old Wives' Tales*, Robert and Michèle Root-Bernstein enumerate the ways in which honey has recently helped heal both external and internal ailments. The Root-Bernsteins report that a bacteriologist tested the antibacterial activity of honey on typhoid, pneumococci, streptococci, dysentery, and other bacteria. "Without exception, all were killed within a few days of exposure to honey, and most within a few hours."

Even present-day eyes have resorted to the remedies of *Ebers* and Dioscorides. Herb Spencer, the president of the Southwest Missouri Beekeepers Association, wrote to the *American Bee Journal* in

April 2004 to describe the benefits of dabbing honey into his ailing eyes. Experiencing blurred vision and excessive tearing when he drove at night, and familiar with the healing history of honey, he

> started putting raw honey in my eye every night. I did this by putting a drop of honey on my forefinger, and pulled my right lower eyelid down and dabbed the honey on my lower eyelid which closed immediately to begin distribution all around my eyeball. It does burn a little bit, but only for a couple minutes. In about three days, I could discern a big improvement, so I started putting it in both eyes ... now I can see to drive at night.

After reading Herb Spencer's testimonial I imitated its author, and the ancients, and tucked a drop of raw honey behind each lower lid when my eyes were feeling dry and fatigued, or, in the words of *Ebers*, when "something bad had happened" to them. (I also swigged an oral dose for good measure.) Spencer was right about the slight burning sensation and also the rejuvenating effects. As the great Columella said of honey, "You will be amazed."

Honey helped my tired eyes and Spencer's night blindness. Some doctors and researchers hope it will tackle some blind spots in areas of conventional medicine. The authors of "Honey—A Remedy Rediscovered" wrote in the *Journal of the Royal Society of Medicine* that "The therapeutic potential of uncontaminated, pure honey is grossly underutilized. It is widely available in most communities, and although the mechanism of action of several of its properties remains obscure and needs further investigation, the time has now come for conventional medicine to lift the blinds off this 'tradi-

A honeyed portion of the *Papyrus Ebers*.

tional remedy' and give it its due recognition." Hippocrates, Dioscorides and other ancient physicians would surely have agreed.

Honey was as popular in the boudoir as it was in the kitchens and pharmacies of history. It is a humectant, which means it promotes the retention of water, so a thin veneer smoothed onto the face or washed onto the body attracts and holds moisture on the skin. Nourishing and antibacterial, the bees' emollient brew has been used to cleanse, soften, and brighten faces from Nefertiti's day to our own. The *Papyrus Ebers* devotes quite a lot of hieroglyphs to cosmetics and honey, suggesting potions for everything from removing spots on the face to growing lustrous hair on the head. "To make the face smooth, take water from the qebu-plant, meal of alabaster, fresh abt-grain, Mix in honey, and anoint the face therewith." Or, "to remedy and beautify the skin, use meal of alabaster, meal of natron, sea-salt, honey, and anoint the body therewith."

Cleopatra, a descendant of Nefertiti, and another notorious beauty, is said to have compiled a book of grooming tips, dedicating many pages to the use of honey. As a proper Egyptian queen, she had great interest in hygiene and scent. Perfuming the body with unguents and oils and thoroughly fumigating one's person and chambers with fragrant incense were part of the daily beauty regimen. The *Ebers*, if she consulted it, had many recipes for such fumigants: "In order to make pleasant the smell of the house or of the clothes, crush, grind and combine myrrh, elderberries, cypress, Resin-of-Aloes, inekuun grain, mastick, and styrax. Put into honey, cook, mix and form into little balls and fumigate with them. It is also worth while to make mouth-pills out of them to make the smell of the mouth agreeable."

Like Cleopatra herself, the Egyptian obsession with fragrance eventually made its way to Rome. Citizens there were soon just as obsessed with banishing wrinkles and smelling good as the Egyptians. Pliny the Elder describes a popular, expensive first-century scent called *metopium*, made with almond-oil from Egypt, to which were added green olive oil, cardamom, rush juice, reed juice, honey, wine, myrrh, balsam seed, and the aromatic resins galbanum and terebinth. These ingredients were exotic and costly, but seduction was worth the expense. "When a girl wearing the best metopium passes, she attracts the attention even of those who should be otherwise occupied."

The poet Ovid, who lived from 43 B.C. to 18 A.D. (during or close to the metopium craze), wrote *The Art of Beauty*, describing the vanities of the day and offering a few lyrical grooming tips:

Learn now in what manner, when sleep has let go your tender limbs, your faces can shine bright and fair. Strip from its covering of chaff the barley which Libyan husbandmen have sent in ships. Let an equal measure of vetch be moistened in ten eggs, but let the skimmed barley weigh two pounds. When this has dried in the blowing breezes, bid the slow she-ass break it on the rough mill-stone: grind therewith too the first horns that fall from a nimble stag. And now when it is mixed with the dusty grain, sift it all straightaway in hollow sieves. Add twelve narcissus bulbs without their skins, and let a strenuous hand pound them on pure marble. Let gum and tuscan seed weigh a sixth part of a pound, and let nine times as much honey go to that. Whoever shall treat her face with such a prescription will shine smoother than her own mirror.

Another poetic potion, claims Ovid, "will clear freckles from the face in a trice, of this about three ounces may suffice, But e'er you use it, rob the labouring bee, to mix the mass and make the parts agree." These were popular antiaging formulas. Pliny reports on a famously vain woman, Polla, who spent her sixtieth birthday immersed in a version of Ovid's recipe, a "poultice of honey and wine lees with finely ground narcissus bulbs."

Poppea, the second wife of emperor Nero, lived a few decades after Ovid and surely was a fan of the honey-narcissus formula. Famously beautiful, conceited, and indulgent, she boasted of having a hundred slaves to keep her thus. Her staff bathed her body in asses' milk and honey twice a day to preserve the luster of her skin and applied a daily facial mask of honey and herbs, washed off also with asses' milk. (This softening, moisturizing bath can presumably be achieved with cow's milk if asses are unavailable.) These efforts, though undoubtedly successful, were not enough to prevent her husband from tiring of her and eventually murdering her. Honey does not cure everything.

Fashionable men and women continued to primp, polish, and perfume themselves with honey for the next two thousand years. In the seventeenth century, *The Queen's Closet Opened*, published in England, offered household and cosmetic hints for ladies, addressing concerns and using ingredients that had not changed from the times of Nefertiti or Poppea. "A water of flowers good for the complexion of ladies. Take the flowers of elder, a flower-de-luce, mallows and beans, with the pulp of melon, honey, and the white of eggs, and let all be distilled...this water is very effectual to take away wrinkles in the face." In a French beauty manual of the mid-eighteenth century, honey was involved in a kind of moisturizing blush stain. "To make a water that tinges the cheeks a beautiful pink: take a quart of white wine vinegar, six ounces of honey, three ounces isinglass; two ounces of bruised nutmegs. Distill with a gentle fire, and add to the distilled water a small quantity of Red Sanders in order to color it." A hundred years later, honey was still a staple for the European beauty. *Toilet Table Talk*, which debuted in 1850 in London, advised ladies to strain and heat a pound of honey, add a pound of white bitter paste, then two pounds of oil of bitter almonds and five egg yolks in alternate portions. Users were instructed to apply the results daily to keep the face fresh, soft, and beautiful.

With the industrialization of the late nineteenth century (and of course the rise of sugar and its sad eclipse of the beehive), beauty and personal care products were no longer made in the home with ingredients gathered from the kitchen, garden, and beehive. Beauty and hygiene increasingly sought rescue in modern science and chemistry, and grooming supplies migrated to the drug or department store. But as in holistic medicine, old-fashioned, natural, and honeyed folk remedies have lately enjoyed a renaissance in beauty. Ingredient lists on many popular products read like an ancient herbal, the contents so natural, effective, and dripping with honey and beeswax that they might have been concocted by Cleopatra herself.

Beeswax is one of the most frequently used ingredients in cosmetics new and old. When wax is emulsified, its plasticity allows it to be spread thinly on the face and body, holding scent, color, moisture, and herbal beautifiers firmly in place like a delicate scaffold or glove. Egyptians used beeswax as a foundation for many creams, including one in *Ebers* that calls for beeswax crushed with incense, fresh olive oil, cypress oil, and fresh milk, then applied for six days to "drive away wrinkles from the face."

Beeswax was also important in Nile Valley hair care, where heads were shaved bare for aesthetic, hygienic, and heat reasons, making wigs a necessity for the pageantry of the temple. Wax was used to sculpt and hold these extravagant fake hairdos in place. Elaborate wigs found in tombs of the late Egyptian dynasties were made of date-palm fiber and grass styled with beeswax. Until the nineteenth century, all over the world, wigs and mustaches were curled, twirled, and set with beeswax. For most of humanity, desirable hairs, such as brows and lashes, could be polished, affixed, and embellished with beeswax, while unwanted ones could be removed with it.

Though it is an ancient art and inquiry, the term "apitherapy" has been employed for the past few decades or so to describe the therapeutic use of hive products. Besides honey and wax, four other bee-related materials have traditionally been applied, or are just beginning to be used, in medicine and cosmetics: pollen, propolis, royal jelly, and bee venom. These compounds and their benefits are still incompletely understood by science, but as the secrets of the bee continue to be revealed, it is clear that the hive has even more magic and beauty in store.

Plant pollen collected by bees has always enjoyed a reputation as a health food. The Egyptians aptly called it "life-giving dust," believing that it was delivered to the hive from the life-giving gods above. Aristotle described the bee as it "carries wax and bee bread round its legs." Bee-collected pollen was often called bee bread, perhaps with the idea that the insect had fetched the loaf from some celestial bakery. The little loaves were more nourishing than bread; they were dense pellets of protein, packed by the bees and garnished with honey. Since most ancient honey was eaten in the comb or crudely strained from it, honey was infused with this life-giving substance.

Stored pollen was also scooped out of the comb, an orange waxy mash of proteins, vitamins, and sugar that could be eaten as a quick energy food. Olympic athletes were said to feast on this ambrosia before the games, and Mayans spread it on tortillas for energy, travel food, and religious ceremonies. The twelfth-century Jewish philosopher Maimonides reportedly used pollen in honey as a sedative and an antiseptic for wounds. His contemporaries in Arabia and China used it as an aphrodisiac and diuretic.

Historically, people got their pollen from the hive, sloppily mixed with honey and wax. In China, special nets were devised and dragged across lakes to harness the pollen fallen on the surface. This valuable commodity was eaten plain, added to other foods and medicinal teas, or combined with honey to make nutritious cakes in times of scarce food supply. In *The World History of Beekeeping and Honey Hunting,* Eva Crane relates the story of an injured U.S. Army officer in China in the 1940s. Locals fed him a mixture of fruit and wind-blown tree pollen and placed a gooey salve of honey and pollen on his injured feet. "The people stored the dry pollen in clay jars and used it as medicine, antiseptic, and food," writes Crane. "They also kneaded honey and pollen together into flat strips which were dried...and eaten daily on hunting trips and during the monsoon season."

Since the 1940s, pollen screens have been used to scrape some of the colorful load of protein from the bees' hind legs as they wriggle into the hive. The harvested miniature loaves are sold in health food stores as a diet supplement, an invigorating source of vegetarian protein. Components vary according to plant sources, but the average pound of pollen has about 100 grams of protein, the same as a pound of sirloin beef. That same amount of pollen has 1,000 milligrams of calcium and 159 of vitamin C, while a steak has 46 and zero milligrams respectively. Pollen contains twice as much potassium as steak and has large multiples more of vitamin A, thiamin, riboflavin, and niacin. It can be eaten plain or as a dietary supplement, or mixed into facial masks and creams as a way to pack these nutrients onto the skin.

In recent years, pollen has proven itself to be more than a food and facial supplement. Plant allergy sufferers believe that ingesting local pollen desensitizes them to the allergens the same way that shots do (the same goes for local honey, which contains trace amounts of the allergens). Bee-collected pollen is also eaten by men who swear that it increases their libido and sexual performance, an as-yet-unproven claim. Proponents have alleged that the bee's protein capsules can cure everything from obesity to cancer and infertility, though scientists have not found any proof to back these assertions.

A couple of years ago, I conducted an unplanned, very unscientific experiment on the effects of eating bee-collected pollen. I was interviewing a beekeeper in Georgia who sold it from a roadside stand to dieters, athletes from the local university, and men seeking a libido boost. All of his customers, he said, swore by its performance-enhancing abilities. When my visit was over, I grabbed a one-pound sample jar to take home with me and began the long drive back to the Savannah airport. As usual, I was short on time and hungry, so I started gulping pollen as if I were eating Grape-Nuts from the box, in big, dry, crunchy, chewy mouthfuls. After eating about a half a cup, I didn't feel hungry anymore. I did feel both elated and sick, stoned and manic, as if I had had about twenty cups of Smiley's coffee. My pupils in the rearview mirror seemed huge, and my hands on the wheel were shaking. Forced to pull into a truck stop to get a drink and calm down, I was sure that two policemen, parked there on a coffee break, eyed me with suspicion. Guzzling water (and driving fast), it took me a couple of hours to metabolize my pollen excess. Now I eat an energizing spoonful in the morning or sprinkle it sparingly on toast, salads, or yogurt. It probably isn't a cure for cancer, but it is potent stuff.

By the first century in Rome, hive watchers had noticed a sticky substance that the bees used to patch, mend, and strengthen their homes. Someone at around this time named it *pro-polis*, meaning "before the city," having observed that bees used it to strengthen and protect their fortresses. Pliny accurately noted that propolis "is obtained from the milder gum of vines and poplars...with it all approaches of cold or damage are blocked." The fragrant cement that the bees collect from the saps and resins of trees is at first sticky and malleable, and is industriously lacquered throughout the interior before hardening into a brittle rust-colored taffy that caulks, weatherproofs, and strengthens the domicile.

Hippocrates wrote of propolis as a medicine respectfully but vaguely, as did Pliny, who reported that it was "of great use for medicaments" and that it was "used by most people as a substitute for *galbanum*" (a widely used astringent resin). Dioscorides describes it more specifically as "the yellow bee-glue that is of a sweet scent... and easy to spread after the fashion of mastick. It is extremely warm, and attractive and drawing out thornes and splinters. Being suffumigated it doth help old coughs and being applied doth take away the lichens. It is found about ye mouthes of hives."

Most bee glue is about 50 percent tree resin, 30 percent wax, 10 percent essential oils, and 10 percent pollen. In addition to keeping the hive airtight and upright, the odor of the tree resins in the propolis acts as a kind of mothball, deterring pestilent invaders from the hive. Tree gums and oils also offer proven antibacterial and antiviral properties, which the bees exploit, spreading a thin layer of this defensive sheen about the hive as if varnishing their home with Lysol. At the entrance of a hive, the application is especially thick, acting as a decontamination zone for bees returning to the hygienic nest.

Propolis was traditionally harvested by scraping it from the hive, which is as laborious and unpleasant as stripping furniture in a swarm of bees. Now plastic screens are placed inside a colony, inspiring the bees to diligently solder the holes full of propolis, which can be harvested when the screens are removed. Though it is not recommended for household caulking, propolis has been found to be useful in medicine. Many of its properties have not been studied, but the superglue of the bees contains benzoic acid, ketone, quercetin, caffeic acid, acacetin, and pinostrobin, all of which have antiviral, antifungal, anti-inflammatory, and antihistamine powers. Anecdotal evidence suggests that the sticky defense of the bees can be used to combat a variety of infections and inflammations of the human fortress.

While pollen and propolis both have ancient and historic medical pedigrees, royal jelly and bee venom have begun to be exploited only in recent decades. Royal jelly is the milky pabulum that worker bees secrete and feed exclusively to a select few fertilized eggs, one of which, on this special diet, will grow into a queen. An elixir of proteins, fats, carbohydrates, fatty acids, amino acids, and vitamins (with a heavy concentration of Bs), this royal food, like many hive products, has been shown to have potent antibacterial and antifungal powers, which may be useful in treating ulcers, indigestion, colds, and infections.

Because of its phenomenal transformative role in the hive, some people have supposed that royal jelly would produce similarly wondrous makeovers in humans. Alas, it works for bee royalty only. Royal jelly packs a load of natural nutrients that strengthen and nourish both body and skin, but it won't transform users into long-living, vastly reproductive queens. Imagine Poppea, Nero's murdered vainglorious wife, getting her hands on a substance that created strong, beautiful, and fertile queens who lived thirty times longer than their mates!

Even the bees' sting produces certain benefits. Charlemagne is said to have crushed live bees onto his aching arthritic limbs, and whole bees, alive and dead, were used frequently in medieval prescriptions. In her *Physica*, Hildegard prescribed "for anyone on whom ganglia grow, or who has had some limb moved from its place, or who has any crushed limbs, take bees that are not alive, but which have died, in a metallic jar. Put a sufficient amount on a linen cloth, and sew it up. Soak this cloth, with the bees sewn within, in olive oil, and place it over the ailing limb. Do this often, and he will be better." A medical dispensary from 1895 describes *Apis Extract*, which was produced by placing a bunch of bees in a jar and shaking it, exciting the frustrated bees to coat the inside with venom. Mixed with alcohol, the resulting liquid was used on infections, rashes, and other inflammations. What Charlemagne and Hildegard were after, though they might not have realized it, was bee venom. It stimulates the adrenal glands to produce cortisol, a powerful anti-inflammatory, which might have soothed the king's arthritis. Apamin, a neurotoxin in bee venom, anesthetizes the sting site, which would have made it numbingly appropriate for Hildegard's patient with a crushed limb.

Bee venom has been commercially harvested only in the last thirty years or so. At first it was squeezed from agitated bees onto an absorbent cloth. More recently a contraption involving an electrified glass plate has been used. When the bees contact the glass near a hive entrance, they receive a mild electrical shock and extend their stingers in alarm, spurting a drop of venom in surprise. Collected from the glass plates, the venom is injected or smeared in creams onto aching joints and arthritic limbs. Those who feel that the sting experience is more stimulating and effective than shots or creams have turned to live bees (purchased from local beekeepers) for their therapy. A carefully applied sting or two numbs the site and stimulates the immune, anti-inflammatory, and circulatory systems, which chain of responses may be useful in combating multiple sclerosis, arthritis, lupus, and chronic fatigue syndrome.

Edmund Hillary, the first man to summit Mount Everest in 1953 at the age of 34, was from a New Zealand beekeeping family. Perhaps his amazing health and fitness came in some part from an inadvertent lifelong program of apitherapy, eating lots of honey and pollen and receiving numerous doses of bee venom. In his autobiography, *Nothing Venture, Nothing Win*, Hillary describes beekeeping and his program of daily stings, which both sound a lot like Smiley's.

> It was a good life—a life of open air and sun and hard physical work. And in its way a life of uncertainty and adventure: a constant fight against the vagaries of the weather. We had 1600 hives of bees spread around the pleasant dairyland south of Auckland, occupying small corners on fifty different farms. We were constantly on the move from site to site. . . . We never knew what our crop would be until the last pound of honey had been taken off the hives; it could range from a massive sixty tons to a miserable twenty or less. But all through the exciting months of the honey flow the dream of a bumper crop would drive us on through long hours of hard labour; manhandling thousands of ninety pound boxes of honey comb for extracting . . . and grimacing at our daily ration of a dozen, or a hundred beestings.

Nearing fifty, Smiley is also in amazingly good shape, which he attributes in part to his honey consumption. "I eat honey six days out of seven," he says, "and I definitely heal faster since I've been working in honey." On the benefits of bee stings, he's less sure. "I've got arthritis, and all those bee stings haven't helped me at all." George, smoking a cigarette as he mends hive frames, wags his hands loosely up and down, and says, "I let 'em sting me on the wrists." He has tendonitis from lifting heavy frames out of their boxes thousands and thousands of times. "I got 'em to sting me as much as possible on my wrists, and it works," he says "Feels much better. Until the numbness wears off." Smiley considers this. He hasn't specifically tried to get the bees to sting him on his back, where the arthritis hurts, or on his wrists, which are sore from beekeeping and, before that, oystering. "I know you have to put them on the right spot. I'll try it," he drawls. "I just don't know if I can take the pain." They both laugh as Smiley slaps a knee.

While the bees are resting in their big medicine ball up north, Smiley and George build hundreds of boxes, wire and repair thousands of frames, sell Christmas honey, medicate, feed, and constantly check on their clustering charges. By January he'll bring them all back down to Wewahitchka to catch the first nectar flows. In February and March he'll start making increases and housing his new livestock. By the middle of April he'll be inspecting tupelo buds, trying to predict when the blossoms will spring forth. And then it all starts again. "This job is never, ever done," says Smiley with fatigued wonder. "Every year's different. You're doing the same things, but every year's different."

Standing in the front yard on a cold winter day, he looks at the new pink house that the bees built for him. He heads toward his work area to build some new homes for them, thinking about all the new bees he'll nurture next year and all the honey and money they'll make for him. "You gotta take care of your bees," he says thoughtfully. "You take care of your bees, and they'll take care of you."

GRATITUDE

While Smiley is taking care of his bees and preparing for 400 new colonies, I'm waiting to see how well my two survive the winter. December still means snow and cold in Connecticut, and lots of it. The bees have been clustering inside their boxes since the first frosts of October, and they'll pretty much stay that way until March or April, when the temperature rises and the scent of forsythia lures them out to investigate. For now, the hives are silent and appear as lifeless as a couple of weathered tombstones. Every time I turn up the thermostat in my house, I wonder how they are doing in theirs.

When I visit the hives in winter, I am met with a disappointing quiet, as if the bees have gone away on vacation. During the spring, summer, and fall, I become accustomed to their welcoming murmur and busy clamor, and I miss it all winter long. Trying to make myself useful, I occasionally clear snowdrifts from the entrances so the bees can get out if they have an opportunity. When the days are warm enough, bees will leave the cluster and fly out to flush the waste they have so graciously been retaining for weeks at a time. (They would sooner die than soil the hive.) On bright sunny winter days, I glimpse them on these exuberant cleansing flights and start to anticipate spring, when I'll be able to see and hear the joyful creatures every day.

Sadly, my initial colony of bees starved and froze to death during that first winter. On a beautiful spring day, I approached the hive greedy and giddy with anticipation as if I were going to reunite with a wonderful old friend. But as soon as I stood next to the box, I knew there would be no reunion. The lively humming charisma of the hives had melted and disappeared with the snow. When I pulled off the cover, an earthy sweet musty sawdust smell wafted up, as if from an old abandoned barn. Lifting a frame out of the top honey super, I saw little bee behinds sticking out of the cells like tiger-eye beads in my childhood Lite Brite. The surrounding storage cells were dry and empty, desperately licked clean of food. Reading the frame like a medical chart or a forensic report, I realized that the bees had starved to death. I had left them with a full super of honey, but it had not been enough to sustain them through the unusually cold, long winter. When the last cold snap had come, the hungry, weakened ball of bees had been flash frozen like a bag of peas.

The frames of the hive presented a dramatic reenactment of the colony's last seconds. As I excavated the colony, I understood how the discoverers of Pompeii might have felt, unearthing layers of life and community frozen in time from the exact moment of their death. Worker bees clung to the comb in mid searching step, and groups dangled together in final desperate clusters. Many had their heads in the comb, feet sticking out, having been diving deep into cells to scrape and suck a few last tastes of honey when they were caught by the cold. Throughout, the storerooms were empty, licked and scavenged clean before the bitter apocalypse. I kept imagining the queen surveying her domain as her children dwindled and died around her. The monarch was nowhere to be found, although I was too upset to sift carefully through the winged mulch of bodies at the bottom of the hive.

In one corner I unearthed a mouse and her litter of almond-sized pink babies, cozily insulated by the pine box, some cattail down, and the delicate shiny husks of dead bees. Dumping the mice and the bee carnage in the woods, I was bereft. In that first summer of hours spent sitting next to the hive and puttering within it, the bees had become my pets, confidantes, and inspiration. They had illuminated the wondrous web of nature around me, given me the miraculous gift of their honey, and I, I thought, had killed them all. I considered becoming a beekeeping dropout. When my sense of guilt and failure abated, I recalled that bees have generously provided and prevailed for centuries, hampered and abetted by nature and man alike. Trials and errors are constant with bees, and that was my final lesson for that first year. Having learned it, I knew I would try again, because I wanted the sweet curriculum of bees, nature, history, and honey to continue being taught in my garden. As Smiley says, "It's hard work to keep bees alive, but it's worth it. They're always teaching you something." Salvaging what frames I could, I called the dealer to order more bees. The next year the winter was shorter and milder and my colonies survived, even without my overzealous offering of an extra super full of honey left on the hives.

Now every winter I look up at the humble hive boxes and contemplate the fierceness and fragility of the bees' lives and the enormity of their effort. I am awed by how much they have taught and given me, and by how much I still have to learn. In winter I dream of honey, of new flowers to plant for the bees, and of other ways to nurture and encourage them when they sally forth.

The love affair that started six years ago is even stronger now than when it began. My sense of wonder, admiration, and respect for the bees has deepened and matured. They have sweetened my life in so many ways. Beekeeping is an endlessly satisfying passion, education, and reward. Looking at the snowy winter hives, I wonder what new lessons, treasures, and flavors the bees will bring me in the spring. I know that if I take care of them, they'll take care of me.

TEN

Some Honey Recipes,
Old and New

The following are adapted from *The Roman Cookery Book: A Critical Translation of* The Art of Cooking *by Apicius*, by Barbara Flower and Elisabeth Rosenbaum. Apicius's cookery was long on honey but short on instruction. Generally, no temperatures, equipment types, or even ingredient amounts are given, so daring culinary intuition is required. At best the results are delicious, at worst they are "interesting" yet historic. If "interesting," add more honey. Or more salt: liquamen and garum were liquid salt sauces of stewed, fermented fish entrails and bones, most similar to the Vietnamese (nuoc nam) and Thai (nam pla) fish sauces found in gourmet and Asian markets. Passum is sweet grape juice, boiled down to half its volume, then further sweetened with honey.

Pisam Vitellianam Sive Fabam
Peas or Beans à la Vitellius (The Roman Emperor in 69 A.D.)

Boil the peas and stir until smooth. Pound black pepper, lovage (or celery leaves), and ginger together, and over this put yolks of hard-boiled eggs, honey, liquamen, wine, and vinegar. Put all this in a saucepan, add oil, and bring to the boil. Season the peas with this. Stir until smooth if lumpy. Add more honey and serve.

Minutal ex Praecoquis
Pork Fricassee with Apricots

Put in the saucepan oil, liquamen, and wine. Chop in dry shallot, add cooked diced shoulder of pork, and sauté. Pound together pepper, cumin, dried mint, and dill, moisten this with honey, liquamen, passum, a little vinegar, and some of the cooking liquor and mix well, then add to the cooked pork. Add the stoned apricots. Bring to a boil, and let it cook until soft. Crumble pastry (bread crumbs can be used) to bind. Sprinkle with pepper and serve.

In Aprum Assum Iura Ferventia Facies Sic
Hot Sauce for Roast Boar

Pound together pepper, grilled cumin, mint, celery seed, thyme, savory, safflower, toasted nuts (pine kernels or almonds), honey, wine, liquamen, vinegar, and a little oil. Cook and serve hot over roast (boar or otherwise).

Embamma in Cervinam Assam
Sauce for Roast Venison

Mix together black pepper, spikenard (or equal parts lavender and valerian), bay leaf, dried onion, celery seed, fresh rue, honey, vinegar, and liquamen; add Jericho dates, raisins, and oil. Boil to create a thick sauce.

Ius in Copadiis
Sauce for Meat Slices

Chop hard-boiled eggs, then add pepper, parsley, cumin, boiled leek, myrtle berries, honey, vinegar, and oil.

Tubera
Truffles

Boil, sprinkle with salt, and put them on skewers and grill lightly. Then in a saucepan combine oil, liquamen, wine, pepper, and honey. When this boils, thicken with cornstarch and serve over the grilled truffles.

Dulcia Domestica
A Homemade Sweet

Stone dates and stuff them with nuts, pine kernels, or ground pepper (or all three). Roll in salt, fry in honey, and serve.

Aliter Dulcia
Another Sweet

Pound pepper, pine kernels, honey, rue, and passum together, then add milk and pastry and thicken while cooking with a few eggs. Pour honey over this custard, sprinkle with pepper, and serve.

---

*The following are adapted from the various authors
and translators of the Loeb Classical Library.*

Savillum
Sweet Cheese Bread
Cato (234–149 B.C.), *On Agriculture*

Half a pound of all-purpose flour, two and half pounds of cheese mixed together. Add a quarter of a pound of honey and one egg. Grease an earthenware dish with olive oil. Mix all the ingredients well, put into the dish, and cover with an earthenware lid. Make sure that you properly bake the middle, where it is deepest. When it is cooked, remove from the oven, drizzle with honey, sprinkle with poppy seeds, put back under the lid briefly to crisp it, and then take it out of the oven. Serve in the warm dish with a spoon.

———————————

Conditurae Olivarum
Olive and Celery Spread
Columella (first century A.D.), *On Agriculture*

Most people finely chop leeks, rue, and mint together with young parsley, and add crushed green olives. Then they add a little peppered vinegar and honey or mead, drizzle with green olive oil, and garnish with green parsley.

———————————

Oxyporium

Salad Dressing

Columella (first century A.D.), *On Agriculture*

Combine one ounce of lovage, two ounces each of skinned raisins and dried mint, and white or black pepper to taste. These ingredients can be mixed with honey.

Cucumbers

Braised Cucumbers

Pliny the Elder (23–79 A.D.), *Natural History*

Peeled and cooked in olive oil, vinegar, and honey, cucumbers are without doubt more delicious.

Elaphos

Sesame Cake

Athenaeus (170–230 A.D.), *The Deipnosophists*

A flat cake in the shape of a deer is served at the festival of Elaphobelia, made from spelt dough, honey, and sesame seeds.

Mykai

Mushrooms in Honey

Athenaeus (170–230 A.D.), *The Deipnosophists*

Mushrooms should be prepared with vinegar, or a mixture of honey and vinegar, or honey and salt alone.

From This Century

Smiley's Marinade for Chicken or Fish

¼ cup honey
Juice of one lemon
Pinch of black pepper
Garlic, minced
Hot sauce (optional)

Combine all the ingredients in a glass or stainless-steel bowl. Brush onto chicken or fish while grilling, or let marinate for a few hours before roasting.

MAKES ABOUT ½ CUP

Smiley's Favorite Smoothie

¼ cup honey
1 cup apple juice
1 cup water
1 banana
3 or 4 strawberries
4 1-inch cubes pineapple

Combine all ingredients in a blender and purée.

MAKES 1 SERVING

Suggestions from the National Honey Board

Honey Mint Chocolate Sauce

1½ cups honey
¼ cup crème de menthe liqueur
1 cup unsweetened cocoa powder

In a small saucepan, combine all the ingredients. Bring to boil over medium-high heat, stirring occasionally; remove from heat. Cool 10 minutes. Serve warm.

MAKES ABOUT 2½ CUPS

Berry Striped Pops

2 cups strawberries
¾ cup honey
6 kiwifruit, peeled and sliced
2 cups peaches, peeled and pitted
12 3-ounce paper cups or Popsicle molds
12 Popsicle sticks

In a blender or food processor, purée the strawberries with ¼ cup of the honey. Divide mixture evenly among the 12 cups or Popsicle molds. Freeze until firm, about 30 minutes. When the strawberry layer is firm, purée the kiwifruit with another ¼ cup of the honey and pour kiwifruit purée into the molds. Insert a Popsicle stick and freeze until firm, about 30 minutes. Purée the peaches with the remaining ¼ cup of honey and pour peach purée into the molds. Freeze until firm and ready to serve.

MAKES 12 POPS

Butternut Squash Soup

2 tablespoons butter
1 onion, chopped
2 cloves garlic, minced
3 carrots, diced
2 stalks celery, diced
1 potato, peeled and diced
1 butternut squash, peeled, seeded and diced
3 cans (14.5 ounces each) chicken broth
½ cup honey
½ teaspoon dried thyme leaves, crushed
Salt and pepper to taste

In large pot, melt the butter over medium heat. Stir in the onions and garlic. Cook and stir until lightly browned, about 5 minutes. Stir in carrots and celery. Cook and stir until tender, about 5 minutes. Stir in the potato, squash, chicken broth, honey, and thyme. Bring the mixture to a boil; reduce heat and simmer 30 to 45 minutes, until vegetables are tender. Remove from the heat and cool slightly. Transfer mixture to blender or food processor; process until smooth. Season to taste with salt and pepper.

MAKES 6 SERVINGS

Sweetly Curried Dipping Sauce

1 cup sour cream
6 tablespoons honey
2 tablespoons cider vinegar
2 teaspoons curry powder
$\frac{1}{2}$ teaspoon ground cumin
$\frac{1}{4}$ teaspoon salt
2 tablespoons chopped cilantro or parsley

In small bowl, combine all the ingredients except the cilantro. Stir until well blended. Cover and chill until ready to serve. Stir in the cilantro. Serve with assorted sliced vegetables.

MAKES ABOUT 1$\frac{1}{2}$ CUPS

Some of My Own Inventions

Blackout Pasta

Use freeze-dried roasted garlic if you can; the crunch in contrast to the pasta is nice.

Salt for the pasta cooking water
Garlic: 1 tablespoon freeze-dried slices soaked briefly in olive oil (optional),
or 2 cloves fresh, very thinly sliced
3 Mediterranean-style oil-cured black olives, shredded
5 teaspoons honey
4 drops sesame oil
4 drops chili oil
Pinch of red pepper flakes
Lots of freshly ground black pepper
4 ounces pasta: linguine, spaghetti, or other long noodle
2 tablespoons heavy cream
4 tablespoons grated Romano or Parmesan cheese

Place a large pot of salted water over high heat and bring to a boil. While the water is heating, if using fresh garlic, gently brown the slices in olive oil until crisp and golden; remove from the oil. In a serving bowl, combine the honey, chili oil, sesame oil, red pepper, and black pepper to taste. Add the garlic slices and olive shreds. Cook the pasta according to package directions. Drain well. Add the pasta to the honey paste and toss it to mix and coat the pasta. Add the cream and cheese gradually, tossing to desired taste and texture. Serve by candlelight. Nudity optional.

MAKES 2 SERVINGS

Honeyburgers

1.5 pounds ground beef
⅓ cup sliced black olives
⅓ cup finely chopped onion
¼ cup capers, drained and rinsed
3 pickled jalapeño peppers, drained and sliced
(add one more if you want really zesty burgers)
2 tablespoons honey
Salt and freshly ground black pepper, to taste

Mix all the ingredients together. Gently shape into 4 or 5 burgers. Cook by your preferred method to your preferred doneness, and serve your preferred way.

MAKES 4 OR 5 BURGERS

Marinade for Fish Steaks or Scallops

¼ cup honey
Fresh gingerroot, the size of a thumb, sliced
Juice of 2 small oranges
Juice of 1 lemon
2 cloves garlic, sliced
Salt and freshly ground black pepper

Thoroughly mix all the ingredients in a shallow, nonreactive container.

MAKES ENOUGH TO MARINATE 2 FISH STEAKS OR 10 LARGE SCALLOPS

Bishop's Barbecue Sauce

½ cup honey
2 teaspoons red pepper flakes
3 cloves garlic, chopped

Mix all the ingredients thoroughly. Use as you would any other barbecue sauce.

MAKES ABOUT ½ CUP

Robbing the Bees Martini

2 teaspoons honey
½ cup really good gin, such as Bombay Sapphire

Mix the honey and gin at room temperature. Pour into a cocktail shaker with ice. Cover and shake vigorously (don't stir!). Strain into 1 or 2 chilled cocktail glasses.

MAKES 1 WHOPPING OR 2 SMALLISH COCKTAILS

Honey Pollen Nut Cake

Butter and flour for the pan
1/2 cup honey
3 eggs, separated
1/2 cup all-purpose flour
2 tablespoons bee-pollen granules
1/3 cup chopped nuts
A dash of ground nutmeg
A dash of ground cinnamon
A handful of currants or raisins, if desired
A pinch of salt

Preheat the oven to 350° F. Butter and flour an 8-inch-square metal baking pan. Beat together the honey and egg yolks until light and fluffy. Mix together the flour and pollen and fold into the honey mixture, then fold in ground nuts, spices, and fruit. Beat the egg whites with a pinch of salt until stiff and shiny. Fold into the batter.

Gently scrape batter into the prepared pan and bake for 30 minutes, or until golden brown.

MAKES ONE 8-INCH CAKE

Beeswax Hand and Foot Balm

*Use a really good olive oil, the kind you would want to have on
your body. The essential oils are available in health food stores.
Lavender and ylang-ylang oils are another great combination.*

3 tablespoons pure beeswax
1 cup olive oil
15 drops bergamot essential oil
15 drops vanilla essential oil

Melt the beeswax in the top of a double boiler over simmering
water. Whisk in the olive oil and 15 drops of each essential oil. When
cooled, it will be the consistency of thick cake icing and can be
worked into the hands and feet. Store at room temperature in a
tightly closed container.

MAKES ABOUT 1 CUP. A LITTLE GOES A LONG WAY.

Honey Pollen Facial Mask

Half a ripe avocado
2 tablespoons honey
2 tablespoons cream
3 tablespoons bee-pollen granules

Let all the ingredients come to room temperature, then blend thoroughly. Clean face and throat and apply the mixture as a mask. Leave on for twenty minutes, then rinse with warm water.

MAKES 1 MASK

Illustration Credits

234 Reprinted with permission from the New York Public
      Library.
241 Reprinted with permission from the Metropolitan Museum
      of Art.

A Final Note

If you are thinking about getting and tending a hive of bees (and you should be), the best place to start is your local beekeeping association, which can be located online. There, you will probably meet an expert (or several) who will likely be more than happy to advise and coach you. Also, read the magazines *Bee Culture* and *American Bee Journal* and one or two of the many practical manuals available. Be prepared to absorb quantities of information and many differing opinions. Then order your bees and begin.

Acknowledgments

With thanks and gratitude to Ace Lichtenstein, who introduced me to bees and beekeeping, and to Maria Massie, who first suggested I write about them and encouraged me throughout. Many thanks also to Mary South, for helping me explore the Florida panhandle and discover the title of the book.

Donald and Paula Smiley put up with my pestering questions for more than two years, were always patient and gracious, and deserve a round of applause.

I have boundless appreciation for my amazing editor, Leslie Meredith, and my agent, Mary Evans, both of whom are bee and honey devotees, and both extraordinary. The copyeditor on this book, Suzanne Fass, is a genius. Michael Burkin, thank you for all of your help, and for leading me to these great women. Thank you also to Martha Levin, Dominick Anfuso, and Kit Frick for making me feel so comfortable and cared for at the Free Press.

Kevin Coyne and Sam Freedman at Columbia taught me so much about reporting and writing, and I thank them, and also the New York Public Library, whose Allen Room offered peace, quiet, stacks of books, and air-conditioning.

Praises and thanks to my exceptional reader, Helen Cassidy Page, and to Mike and Liz Bishop, who offered great suggestions and advice.

And, in no particular order, with apologies to anyone I forgot, thank you so much to the friends and family who helped me survive two years in the wilds of the library and the Florida panhandle: Polly, Heidi, Saba, Sarah, Dave, Riley, Bob, Cindy, Caitlin, Rebecca, Andre, Kerry, Griffin, Lisa, Katherine, Jeff, Elizabeth, Bill, Mark, Judy, Jackie, and Briar. Trix Rosen, thank you for making everything look good.

INDEX

Page numbers in *italics* refer to illustrations.